Henry Coddington Meyer

Water-Waste Prevention

Its Importance and the Evils Due to its Neglect

Henry Coddington Meyer

Water-Waste Prevention
Its Importance and the Evils Due to its Neglect

ISBN/EAN: 9783337139100

Printed in Europe, USA, Canada, Australia, Japan

Cover: Foto ©berggeist007 / pixelio.de

More available books at **www.hansebooks.com**

WATER-WASTE PREVENTION:

ITS IMPORTANCE

AND

THE EVILS DUE TO ITS NEGLECT.

WITH AN ACCOUNT OF THE METHODS ADOPTED IN VARIOUS CITIES IN GREAT BRITAIN AND THE UNITED STATES.

BY

HENRY C. MEYER,

EDITOR OF THE SANITARY ENGINEER.

WITH AN APPENDIX.

NEW YORK:
THE SANITARY ENGINEER.
1885.

Copyright, 1885

By THE SANITARY ENGINEER.

THE SANITARY ENGINEER PRESS,
140 WILLIAM STREET, N. Y.

PREFACE.

DURING the summer of 1882 the Editor of THE SANITARY ENGINEER carefully investigated the methods employed in various cities in Great Britain for curtailing the waste of water without subjecting the respective communities to either inconvenience or a limited allowance. The results of this investigation appeared in a series of articles entitled "New York's Water-Supply," the purpose being to present to the readers of THE SANITARY ENGINEER such facts as would stimulate public sentiment in support of the enforcement of measures tending to prevent the excessive waste of water so prevalent in American cities, and especially the city of New York, which was then suffering from a short supply.

Numerous requests for information, together with the recent popular agitation in connection with a proposition to increase the powers of the Water Department of New York City with a view to enabling it to restrict the waste of water, have suggested the desirability of reprinting these articles in a more convenient and accessible form, with data giving the results of efforts in this direction in American cities since the articles first appeared, so far as they have come to the author's notice; and also to furnish information regarding the enforcement of needed regulations to which there might be more or less opposition due to misapprehension. A careful perusal of this record of the experience of other cities will indicate that it is entirely feasible to control and improve the character of the water-fittings and plumbing-work without requiring the use of patented appliances, and that no special system for all places is recommended.

If the publication of this little book increases the popular appreciation of an urgent public necessity, and tends to secure for water-works officials who are striving to do their duty in the matter of restricting the waste of water, the support of influential members of their community, the author will feel amply repaid for the labor involved in gathering and presenting this information.

NEW YORK, May, 1885.   H. C. M.

TABLE OF CONTENTS.

            PAGES

CHAPTER I.—CONDITION OF NEW YORK'S WATER-SUPPLY.

    Mr. Thomas Hawksley on Advantages of Waste-Prevention—Condition of Water-Supply in England Thirty Years Ago—Means Adopted to Prevent Waste in Great Britain—Norwich First City in England to Adopt Measures of Prevention—London, the Practice There.. 7–10

CHAPTER II.—GLASGOW.

    District Meters Tried as an Experiment—Results of Experiments—Prevalence of Defective Fittings—Testing and Stamping of Fittings—Rules Governing Plumbers' Work................................... 11–14

CHAPTER III.—MANCHESTER.

    History of Waste-Prevention Measures—Methods of House-to-House Inspection—Duties of Inspectors—Methods of Testing and Stamping Fittings.. 15–18

CHAPTER IV.—LIVERPOOL.

    Change from Intermittent to Constant Supply—Method of Ascertaining Locality of Waste by Use of District Meters—Method of House Inspection—Method of Testing Fittings....................... 19–24

CHAPTER V.—PROVIDENCE AND CINCINNATI.

    Review of Measures to Prevent Water-Waste in the United States prior to 1882—Providence, R. I., Results following the General Use of Meters—Cincinnati, Methods of House Inspection with the aid of the Waterphone—Results Attained...................................... 25–30

CHAPTER VI.—NEW YORK.

    Measures Adopted by the Department of Public Works prior to 1882... 31–33

CHAPTER VII.—GENERAL CONCLUSIONS.

Points to be Considered in Adopting Measures for Large Cities...... 34–37

APPENDIX.—POINTS SUGGESTED IN THE CONSIDERATION OF VARIOUS METHODS.

Water-Waste Prevention in Boston in 1883 and 1884—Results attained... 39–46

Waste-Prevention in New York City............................ 46–48

Liverpool Corporation Water-Works Regulations................ 49–51

Glasgow Corporation Water-Works Regulations—Description of Standard Fittings—Penalties for Violations........................ 51–57

Cistern *versus* Valve Supply to Water-Closets in New York City—New York Board of Health Regulations concerning Water-Supply to Water-Closets.. 57–70

Letters from Water-Works Authorities sustaining the action of the New York Board of Health in requiring Cistern Supply to Water-Closets.... 58–61

Extracts from Report of Boston City Engineer on Wasteful Water-Closets... 63–64

Proposed Water-Rates on Water-Closets in New York.............. 64–65

Resolutions of the New York Board of Health endorsing the proposed Water-Rates for Water-Closets.................................. 66

Excerpts from Articles explaining Methods of Arranging Water-Supply to Water-Closets to secure the Minimum Water-Rate in New York (with Illustrations).. 66–70

TABLE OF CONTENTS.

PAGES

CHAPTER I.—CONDITION OF NEW YORK'S WATER-SUPPLY.

Mr. Thomas Hawksley on Advantages of Waste-Prevention—Condition of Water-Supply in England Thirty Years Ago—Means Adopted to Prevent Waste in Great Britain—Norwich First City in England to Adopt Measures of Prevention—London, the Practice There.. 7–10

CHAPTER II.—GLASGOW.

District Meters Tried as an Experiment—Results of Experiments—Prevalence of Defective Fittings—Testing and Stamping of Fittings—Rules Governing Plumbers' Work................................. 11–14

CHAPTER III.—MANCHESTER.

History of Waste-Prevention Measures—Methods of House-to-House Inspection—Duties of Inspectors—Methods of Testing and Stamping Fittings... 15–18

CHAPTER IV.—LIVERPOOL.

Change from Intermittent to Constant Supply—Method of Ascertaining Locality of Waste by Use of District Meters—Method of House Inspection—Method of Testing Fittings..................... 19–24

CHAPTER V.—PROVIDENCE AND CINCINNATI.

Review of Measures to Prevent Water-Waste in the United States prior to 1882—Providence, R. I., Results following the General Use of Meters—Cincinnati, Methods of House Inspection with the aid of the Waterphone—Results Attained.. 25–30

CHAPTER VI.—NEW YORK.

Measures Adopted by the Department of Public Works prior to 1882.. 31–33

CHAPTER VII.—GENERAL CONCLUSIONS.

 Points to be Considered in Adopting Measures for Large Cities......   34–37

APPENDIX.—POINTS SUGGESTED IN THE CONSIDERATION OF VARIOUS METHODS.

 Water-Waste Prevention in Boston in 1883 and 1884—Results attained............   39–46

 Waste-Prevention in New York City............   46–48

 Liverpool Corporation Water-Works Regulations.............   49–51

 Glasgow Corporation Water-Works Regulations—Description of Standard Fittings—Penalties for Violations............   51–57

 Cistern *versus* Valve Supply to Water-Closets in New York City—New York Board of Health Regulations concerning Water-Supply to Water-Closets............   57–70

 Letters from Water-Works Authorities sustaining the action of the New York Board of Health in requiring Cistern Supply to Water-Closets....   58–61

 Extracts from Report of Boston City Engineer on Wasteful Water-Closets............   63–64

 Proposed Water-Rates on Water-Closets in New York..............   64–65

 Resolutions of the New York Board of Health endorsing the proposed Water-Rates for Water-Closets............   66

 Excerpts from Articles explaining Methods of Arranging Water-Supply to Water-Closets to secure the Minimum Water-Rate in New York (with Illustrations)............   66–70

CHAPTER I.

CONDITION OF NEW YORK'S WATER-SUPPLY.

MR. THOMAS HAWKSLEY ON ADVANTAGES OF WASTE-PREVENTION—CONDITION OF WATER-SUPPLY IN ENGLAND THIRTY YEARS AGO—MEANS ADOPTED TO PREVENT WASTE IN GREAT BRITAIN—NORWICH FIRST CITY IN ENGLAND TO ADOPT MEASURES OF PREVENTION—LONDON, THE PRACTICE THERE.

THE condition of New York with reference to its water-supply has frequently been referred to in THE SANITARY ENGINEER during the past three years,[*] and the situation of Boston, Philadelphia, and Cincinnati, to mention no smaller places, has suggested the propriety of reviewing the experiences of other cities in their attempts to husband their resources in the matter of water-supply. We have for some years looked into the question of water-waste prevention, and during the past summer have spent considerable time in investigating the methods adopted in the cities of Great Britain, with a view to being able to form an intelligent opinion as to the best practicable methods that might be employed in New York, making due allowance for the great difference in local conditions, as to the habits and requirements of water consumers and the methods of municipal administration. Briefly stated, the condition of New York is as follows: The Croton aqueduct began delivering water to this city to its fullest safe capacity ten years ago. That capacity was about ninety-five million gallons daily. To furnish this amount of water the flow-line in the aqueduct has been maintained at a point nearly two feet higher than the designers and constructors of the aqueduct anticipated when it was built. The strain occasioned thereby has, therefore, been a matter of deep solicitude on the part of the engineers whose duty has been the maintenance of the conduit.

During the past ten years the population of this city has increased about three hundred and ten thousand. Manufactures, hydraulic-

---

[*] Written in 1882.

elevators, and other demands upon the limited amount of water have multiplied at an enormous rate, and no doubt faster than all has the waste increased.

Besides all this, during the fall of 1881 it was made painfully apparent that the storage capacity of the Croton reservoir was inadequate, and while in ordinary times the flow over the dam into the Hudson averaged during the past eighteen years three hundred thousand gallons daily, yet this average was of no avail in a period of extended drought. From the opinions of the ablest hydraulic engineers in this country it is evident that the present Croton water-shed is ample for the present and future needs of this city, provided, however, that additional storage capacity be arranged for and an additional conduit be provided to carry the water to this city.

As a matter of immediate concern, the following articles will be confined to the description of methods by which present and future suffering can be mitigated, and that is nothing more or less than the prevention of waste, for even should the work of building a new aqueduct be begun at once, it will probably be seven years before an additional gallon of water can be delivered through it ; meanwhile the probabilities of suffering are not pleasant to contemplate.

The consumption of water in New York, with the present reduced pressure in the mains, is about seventy-five gallons daily for each inhabitant, which amount is nearly three times as great as Liverpool and Manchester use with their greatly increased pressure. There is no doubt that, after making the most liberal allowance for the difference in the habits and usage of our people, fully forty gallons per head are wasted, and of this amount twenty-five gallons at least may be considered as preventable waste.

At the ratio of increase of population during the past ten years, another seven years will add two hundred thousand to our numbers, which means a decrease of nearly one-fifth in the water-pressure now maintained.

It is obvious, then, that in order to supply the demands of this increased population and retain the present inadequate pressure, one-fifth of the amount delivered, or about twenty million gallons daily, must be saved from the amount now wasted, and it is a matter much to be regretted that so little has yet been done in that direction.

As an illustration of what may be accomplished under proper regulation to suppress water-waste, we quote from an address of Mr. Thomas Hawksley, past President of the Institution of Civil Engineers, delivered at Liverpool, in 1876 :

"The necessity for the adoption of a uniform and uniformly applied system of rules and regulations is manifested by the results of their application or otherwise in towns of a generally similar character. Oxford and Cambridge are two such towns. In the former, where no efficient regulations are in use, the quantity delivered is eighty gallons per head; in the latter, where efficient regulations are in use, the quantity delivered is eighteen gallons."

Thirty years ago the water-supply of nearly every English city was on the intermittent plan, it being supposed that under a constant service the waste would be so great that the water companies would be unable to keep up the supply. In cities, therefore, where a scarcity of water was imminent, and with very little prospect of an early increase, attention was turned, as it never is in times of plenty, to the adoption of measures for the suppression of waste.

The means adopted in Great Britain to check waste are substantially three:

First—Intermittent supply, and an efficient inspection and supervision over fittings and plumbing-work.

Second—Testing and stamping of all fittings, rigid rules to govern the plumbing-work, inspection before water is turned on, and house-to-house inspection periodically thereafter.

Third—The Liverpool plan, which is to have district meters in the streets in order to localize the waste, rigid and persistent inspection, careful testing and stamping of every fitting used, and inspection of the plumbing-work periodically thereafter.

Norwich was the first city in England to secure parliamentary powers to adopt and enforce rules controlling the character of the water-fittings and the plumbing. This was made necessary, as the consumption had so increased that the intermittent supply had to be reverted to. The consumption had then risen to what in England is regarded the enormous amount of forty imperial gallons per head daily—about forty-eight United States gallons. Under the direction of the eminent hydraulic engineer, Mr. Thomas Hawksley, parliamentary powers were sought and obtained, rules were prepared and enforced, defective mains replaced, and waste stopped to such an extent, that the consumption was brought to fifteen gallons per head—less than one-third the former amount—and the constant service was restored and has ever since been maintained.

Many other cities sought and secured the power to regulate the use of water-supply fittings and control the manner of doing the plumbing-work, and the essential features of the Norwich Water-Works rules are embodied in every set of water-works regulations in Great Britain, while as a matter of fact hardly a single one of them obtains in any American city.

It must not be supposed that the powers to regulate these matters were easily obtained ; on the contrary, they were fought inch by inch, and were only secured by most persistent effort and the expenditure of large sums of money by the various companies. As an instance, we would cite the case of the adoption of the rules to govern fittings and methods of plumbing in Sheffield. The hearing before the magistrates, who by the act of Parliament were required to sanction them, continued thirteen days and cost the water company twenty thousand dollars.

In London the water is supplied entirely by private companies, and with the exception of one or two the supply is yet intermittent. Being private corporations, they have not been able to secure such control over the fittings and the plumbing-work as some of the other cities possess, due, probably, to the fact that powers of the kind sought are not readily granted to any but municipal authorities. The rules under which they now act were adopted about 1873, and required the sanction of the Board of Trade, who appointed as commissioners, to confer with the water-works' managers, the Duke of Richmond, Captain Tyler, and Mr. Robert Rawlinson.

The original demands of the water companies were considered excessive and were strongly resisted. They were modified by the Board of Trade, and finally agreed to. Being private corporations, the feeling against the water companies is much like that against gas companies. The managers, therefore, in order to prevent controversies, and relying on their intermittent service to restrict consumption, are more lax in the enforcement of their published rules than are the authorities of Liverpool, Manchester, and Glasgow. The result is that many so-called waste-preventing devices are used in London, and winked at, that are not permitted in the other places, the effect of which, however, is to afford a market for the inventors of new things.

CHAPTER II.

GLASGOW.

DISTRICT METERS TRIED AS AN EXPERIMENT—RESULTS OF EXPERIMENTS—PREVALENCE OF DEFECTIVE FITTINGS—TESTING AND STAMPING OF FITTINGS—RULES GOVERNING PLUMBERS' WORK.

GLASGOW is favored with what is now considered an abundant water-supply, and this happy condition is due to the successful effort of Mr. Gale, the Water Engineer, to suppress waste. The practice is to require plumbing fittings to conform to a standard set by the Chief Engineer, to have the plumbing of each house examined before water is turned on, and periodically thereafter.

Some years since, when Glasgow was threatened with a scarcity of water, increased powers and funds were allowed the Chief Engineer to secure the suppression of waste, and the results are stated at length in Stewart's paper on "Prevention of Waste of Water."

In certain localities, district meters were applied by way of experiment. These showed that it was possible to reduce an average consumption of 59.2 gallons (imperial) per head per twenty-four hours to 26.6 gallons. The night rates, which had been 45.0, were reduced to 10.8.

These gratifying results secured the extension of this system to forty-one sub-districts, and we quote the following from the paper read by Mr. Thomas Stewart at the Institution of Civil Engineers:

"A few of the more striking results are given here:

NUMBER OF DISTRICTS.	AT STARTING OF METERS.		AFTER THE THREE FIRST INSPECTIONS.	
	Total. Gallons.	Night rate per 24 hours. Gallons.	Total. Gallons.	Night rate per 24 hours. Gallons.
I. 2............	81.8	64.0	34.1	9.9
I. 3............	68.3	45.7	31.9	15.0
II. 2	over 70.0	over 70.0	35.5	17.5
IV.2............	41.6	27.4	27.4	8.3
IV.6............	61.0	55.0	47.2	23.8

"In seventeen districts, at the starting of the meters, the consumption per head per twenty-four hours was 60 gallons, or more; in eleven of these it was reduced by the first three inspections below 40 gallons. In twenty-five districts at the starting of the meters, the night rate per twenty-four hours exceeded 40 gallons; it was reduced below 20 gallons by the first three inspections.

"Appendix III. shows that, at the starting of the meters, the average consumption for all purposes was 49 gallons per head per twenty-four hours, which was reduced to 32 gallons by the first three inspections; the corresponding night rates were 37.7 and 17.5 gallons. The night rates, however, are exaggerated, on account of the pressure being much higher at night than during the day. On the 6th of January, 1881, the consumption was 35.3 gallons, which shows a saving of 13.7 gallons per head; or, if the whole district under control be considered, the saving is 1,114,317 gallons per twenty-four hours, equal to 1.48 gallon per head of the whole population in the area of supply. If a saving of 13.7 gallons per head was made over the whole of Glasgow and district, containing 754,778 persons, the total saving would be 10,340,450 gallons per twenty-four hours.

"The operations showed, almost uniformly, that the quantity of water consumed increased during the nine weeks which elapsed between two inspections, and the most reliable observations gave an increase of 3.4 gallons per head, on an average. The total saving per head after the first three inspections were completed was 17 gallons, so that if the consumption continued to increase at the rate of 3.4 gallons in nine weeks, the original rate of consumption would be reached in ten months and a half. In some of the districts the increase was at the rate of only 2 gallons in nine weeks; in others it was at the rate of 7 gallons.

"An instance of the gradual increase in the rate of consumption may be referred to in sub-district IV. 3, when left without inspection for three months and a half. Before the night inspection the consumption per head per twenty-four hours was 27.0 gallons and the night rate 15.3 gallons. Fourteen days after the inspection had been made the rates were 22.2 and 10.1 gallons; six weeks after the inspection the rates were 23.3 and 9.6 gallons; ten weeks after the inspection the rates were 26.5 and 12.6 gallons; and fourteen weeks after they had increased to 28.0 and 14.9 gallons."

It will be noticed that the application of district meters was not general in Glasgow, though the results obtained by their partial use no doubt secured for the Chief Engineer the authority and means to enforce a rigid house-to-house inspection, and to control the character of new fittings, though, as a matter of fact, more latitude is allowed in the trial of new taps and devices in Glasgow than in Manchester, Liverpool, and some other places. As may be supposed, manufacturers and inventors are constantly submitting new, or alleged new, devices for the approval of the engineer, claiming, as a matter of course, that they are superior to anything heretofore in use. Many are rejected on sight; others that hold out a prospect of success the engineer disposes of by consenting to allow them to be used in a certain house or row of houses if the builder requests it. A record is kept of such cases and their operation is watched for a year. As a general thing few articles survive this test, even of those that are subjected to it, and the authorities find it an easier way of disposing of sanguine inventors than to absolutely refuse to try their wares.

In 1875 Mr. Gale, the Water Engineer of Glasgow, stated that, in that city, as the result of many years' inspection, he found one tap wasting water for every eight taps examined. As the average number of taps in a house was three, he held that every third house was wasting water. Following this line of reasoning, about every house in this city is wasting water, more or less, during the twenty-four hours, since, instead of three taps per house, in New York the average is not less than twelve.

He very strongly urged the improvement of defective fittings, and after admitting the advantages of district meters and appliances for indicating where waste was going on, he yet held that the removal of imperfect fittings was the only possible solution of this difficult question, and "he did not anticipate any improvement in this direction until the authorities instituted a system of testing and stamping every fitting before it was allowed to be used, which had been so long successfully carried out in Manchester." It will be remembered this was stated prior to the enforcement of the rules which secured the gratifying results reported in our last article. In referring thus at length to Glasgow, we do so because our condition now is much the same as theirs was in 1875, and the remarks of Mr. Gale apply with special force to us to-day, especially when he says: "In Glasgow the water-fittings were very numerous, and as the great bulk of the house property had been put up by speculative builders, they were, as a rule, neither sufficiently strong nor properly finished." How absolutely true this statement is, if applied to this city, every plumber and dealer in plumbers' supplies knows.

The rules governing the class of fittings allowed, and the methods of doing the plumbing-work, were printed at length in THE SANITARY ENGINEER of May 25, 1882, page 538; * the leading features of which not in force in this city are the following :

---

* See Appendix.

"A stop-cock on the supply from street-main to house.

"Definite rules to govern the construction of cisterns with reference to the manner of attaching the ball-cock.

"The location of cisterns, the location of the overflow-pipe in a cistern with reference to the water-line when ball-cock is closed, also the terminus of the overflow-pipe.

"No fitting or appliance is allowed to be used that does not bear the maker's name and conform to a standard set by the chief engineer, and is not able to stand the tests to which it is subjected by the testing officers.

"Ground key bibbs, or taps, are not allowed on pipes direct from the main.

"No water-closet can be supplied direct from the main, nor from a tap of any kind. Every water-closet must be supplied from a special cistern so constructed as not to be capable of discharging more than a given amount of water at each flush, and so that it cannot be made to flow continuously, either by intention or neglect."

The clause of the act under which the Glasgow rules are enforced is as follows:

"SECTION 17. All the apparatus used or to be used for conveying water to the houses of the inhabitants and manufactories, or other premises supplied or to be supplied with water under the provisions of the recited act, shall be subject to the approval of the engineer to the commissioners; and in case of dispute between the parties providing such apparatus and such engineer, such dispute shall be determined by the Water Committee of the Commissioners, whose decision shall be final."

CHAPTER III.

MANCHESTER.

HISTORY OF WASTE-PREVENTION MEASURES—METHODS OF HOUSE-TO-HOUSE INSPECTION—DUTIES OF INSPECTORS—METHODS OF TESTING AND STAMPING FITTINGS.

In 1851, Manchester, Eng., received for the first time an abundant supply of water from its new water-works, and has since been able to maintain a constant service. The managers, realizing the importance of controlling the character of the plumbing-work and fittings, attempted, like the authorities of Norwich, to stop the waste, acting under the authority conferred by the water-works clauses act of 1847, but it was found inadequate and they were unsuccessful. In 1858, 1860, and 1867, however, the corporation secured additional authority from Parliament, and has since effectually controlled these matters. Manchester is therefore to-day a city where the testing of fittings and constant and systematic inspection of plumbing is solely relied on to restrict the waste of water.

We believe an average pressure of upward of forty pounds to the inch is maintained, and yet under the admirable management there prevailing the amount of water delivered for all purposes is only about twenty imperial gallons daily per capita (about twenty-four U. S. gallons). Thirteen of this amount is used for domestic purposes and seven for manufacturing, street-watering, and fires. The authorities also claim that they maintain a sufficient pressure to be able to put out fires in the highest buildings without the aid of fire-engines. It should be stated, however, that water-closets are not universal in Manchester, the tub system for the removal of excreta being relied on by a portion of the population. They use meters in all buildings except dwellings, and exempt them on sanitary grounds; where meters are used an ample supply for flushing water-closets is insisted on. The meters used are all tested, and it is claimed that the variation does not exceed two per cent.

Mr. T. H. G. Berry, the superintendent, to whom we are indebted for much information, has stated that he preferred house-to-house inspection, because it enabled the authorities to keep informed as to the use to which water, not otherwise reported, was put, and the increased revenue from these discoveries amounted to about £1,200 a year, which, with the fees received for stamping fittings, practically paid the cost of the entire inspecting and testing staff.

Plumbers are supposed to do the work in a building in accordance with the prescribed rules. When a building is fitted up, and it is desired to have the water turned on, the plumber sends in a form filled out describing the character and number of fittings used. An inspector is then sent to see if they are all stamped and the work done in accordance with the regulations; if so, he makes a report to that effect. If not satisfactory the inspector writes to the plumber and tells him what is needed to secure his approval of the work, and the matter rests in the inspector's hands until his demands are complied with. The good faith of this inspector is tested by a chief or second inspector, who has the first inspector's reports to verify.

The second inspector is in a different office from the first. He takes the reports for verification from the books in that office. If the first inspector is satisfied that the fittings and work in a building are all right, he makes an agreement with the owner as to the water-rate to be paid, it being his business to find the owner and get the agreement signed if the building is to be then occupied. Quarterly rentals are charged to the tenant and weekly rentals to the landlord. It will be obvious that, from the responsibility placed on these inspectors, they must be men of integrity and intelligence—a combination, we regret to say, not always found in men too often selected for such positions in this city.

There are eight so-called first or plumbing inspectors, who make these agreements; and there are two second or supervising inspectors, who only visit the building after the water is turned on, to see if any change has been made, if waste exists, and if the proper rental is being paid. If an extra water-closet is put up or any change made, and not reported to the water-office, the offending plumber is brought before the Water Committee, and either fined or his name stricken off the list.

Of the waste-preventing staff for house-to-house inspection there is one regular member, and the others are used as time from new work permits.

Having explained the details of the house-to-house inspection of water fittings as practiced in Manchester, we will now describe the methods adopted in the examination, testing, and stamping of fittings.

As before stated, meters are used in all buildings except dwellings. The following is the plan adopted in the testing of the meters:

There are two tanks, one holding 110 gallons and the other 180 gallons. On the front of the tank is a glass tube like a thermometer-glass; the water-line indicates on this tube the exact quantity in the tank. The meter is first tried by letting a hundred gallons run through as quickly as possible. The other test is to allow it to pass through in a period of ten hours. As the quantity passed by the meter in either case is known, the reliability of its record can be ascertained.

It has been stated that the receipts from fees for stamping fittings, together with the additional rates collected for uses to which water is put, discovered only by the house-to-house inspection, pay the entire cost of maintaining the waste preventing staff. The following scale of charges may, therefore, be of interest: For bibbs and stop-cocks, twopence each; ball-cocks, threepence; cisterns, sixpence; water-closets, sixpence.

Three men do all the testing in Manchester, and they approve and stamp on an average about fifty thousand fittings in a year, which includes bibbs, stops, cisterns, water-closets, stop-cock valves, and any fittings used in the distribution of water. A portion of these stamped fittings are used in other towns in England, the practice being that smaller places that cannot afford to maintain a testing office require that fittings used in connection with their water-supply shall bear the Manchester, Liverpool, or Glasgow stamp, or that of any other city whose stamp is a guarantee that the article is reliable. By this means it will be seen that other places contribute something toward the support of the testing office.

The rejection of fittings at the present time, after, of course, a number of years' inspection, is about as follows: the four leading makers, about five per cent., the others about twelve per cent. Most of the rejections are due to imperfect seatings, it being understood that compression or screw-down cocks are required; sand-holes, also, are a very considerable cause of rejection. The former defect is revealed by taking the tap apart; the latter by the pressure test, which is 300 pounds to the inch.

Ball-cocks have the size of the ball and length of the lever prescribed. They are tested by turning them upside down and placing a half-pound weight on the ball, and then subjected to a hydraulic pressure of 150 pounds to the inch, which is regarded as equivalent to three hundred pounds when the ball is floating. The lever is stamped at both ends to prevent a plumber from shortening it. Waste-preventing cisterns are carefully examined, and care is taken to see that the valves are so hung that they cannot both be suspended or off their seats at the same time.

The fittings are usually sent by the dealers in quantities to the testing office, as the plumbers will not buy them unless stamped, except in the case of an odd fitting required for special work, which the plumber, to save time, sends direct.

Taps are first weighed singly. If of requisite weight, they are then taken to pieces, and a gauge is used to ascertain that the seat allows a good bearing for the washer. It is also examined to see that the thread is sufficiently long so that the stem will fit well in the cap and not wobble. Any deficiency in either of these particulars secures the rejection of the tap. The stamping of the maker's name on the tap is not insisted on, and with well-recognized makers every tap is not taken apart.

Doubtless the attempt to enforce similar rules in this city would provoke strong remonstrances from dealers and plumbers, who might at times feel annoyed at the restrictions. The former, especially, would have to find other markets for inferior goods and undoubtedly would make the strongest opposition. In this connection, therefore, it may be well to recall the statement of a leading plumber in Manchester, who, when asked his opinion of the rules and their working, replied that through oversight of his workman, and failure to report changes made, he had several times been fined; that at first he had felt much annoyed at being obliged to wait to get special fittings stamped, yet, after his years of experience, he was bound to admit that his punishment had been deserved, was just and proper, and that were he a water-works engineer, intrusted with the duty of serving the public by preventing the waste of their water, he should do just as the authorities of Manchester did.

CHAPTER IV.

LIVERPOOL.

CHANGE FROM INTERMITTENT TO CONSTANT SUPPLY — METHOD OF ASCERTAINING LOCALITY OF WASTE BY USE OF DISTRICT METERS — METHOD OF HOUSE INSPECTION.

WE have seen that in Manchester the testing of fittings and systematic house-to-house inspection is relied on to prevent the waste of water. Liverpool, however, is a city where remarkable results have been obtained by the use of district meters in addition to the measures employed in Manchester. A full description of the district meter system as applied and carried out in Liverpool we gave in a series of letters on pages 494, 552, and 586 of Vol. IV.*

In the present article, therefore, we shall simply recall the leading facts.

Prior to 1873 Liverpool had an intermittent water-supply, the daily consumption being about twenty-nine imperial gallons per capita (nearly 35 U. S. gallons). When constant service was given the amount delivered ran up to thirty-eight gallons. The then limited amount of water accessible demanded either that constant service should be abandoned and the miserable intermittent supply reverted to, or that the waste should be stopped. In this emergency Mr. G. F. Deacon, the Chief Engineer, devised a meter to be placed on the street-mains in the various districts of the city, the main purpose being to indicate whether waste was occurring or not, and if so, whether in buildings or in the mains.

It will be understood, then, that the district meters were relied on to indicate where waste was occurring, and only such places were examined; in other words, the citizens of Liverpool are not subjected to house-to-house inspection, a building not being visited unless the inspectors, by the outside means referred to, ascertain that waste is going on within.

Plumbing-work, however, has to be done in accordance with prescribed rules, and the standards adopted for the various kinds of plumb-

---

* See abstract in Appendix.

ing fittings, together with the system of examination and testing, is not surpassed anywhere.

This is a brief outline of the situation in Liverpool, but those of our readers interested will find a very full description by the writer in the fourth volume of THE SANITARY ENGINEER. The following extract, however, gives an idea of the method of utilizing the district meters :

"The Liverpool plan divides the city into districts. In each district a water-meter is placed in connection with the main. The meter is put under the sidewalk, or in the street, as near the main as possible, and its connection with the main and the arrangement of the pipes supplying the district are such that all the water used in the district must pass through the meter.

"By examining the meters for several successive days the districts in which the greatest waste occurs are ascertained, and the inspectors are able to concentrate their energies on these. The presence of waste is indicated by the peculiar character of the lines on the diagrams taken from the drums of the meters. When waste alone is taking place, as from 2 A. M. to 5 A. M., when, of course, ordinary consumption is not taking place, the flow of water is uniform, and this condition is indicated on the diagrams by a comparatively regular horizontal line, whereas, when water is being drawn off for use, the rate of flow must evidently be variable, and this is indicated by irregular vertical lines.

"There are four gangs of inspectors, each consisting of one chief inspector and one or two assistants. From diagrams brought to the office each morning by the meter man whose duty it is to bring them in, four of the worst districts are selected, and a gang appointed for each, As there are 212 districts in Liverpool, if the inspections were regular it would require about fifty days to get around. Diagrams are, however, brought in each morning from only forty or fifty districts, and from these the selection is made. The average population of a district is about 3,500, but smaller districts are preferable.

"The inspection of the districts selected is made at night, and the details of the inspection are as follows : After 11 P. M. a night inspector visits the most wasteful districts. He has no access to the meter, which is locked. At each outside stop-cock, one of which is placed under the footway at each house or block of premises, he uses a wooden rod or his stop-cock bar as a stethescope. Probably at one out of every eight or ten stop-cocks he hears the sound of running water. In each such case he closes the stop-cock and enters in his note-book the name of the street, the number of the premises, and the time, and makes a chalk-mark on the footway across the stop-cock. Having completed the district, he retraces his steps and reopens the stop-cocks marked with chalk. He then returns to the night office, writes out his notes in copying-ink on the left-hand sheet of a book, and between 6 and 8 A. M. his work for the twenty-four hours is finished. At 9 on the same morning a press copy is taken from the night inspector's report.

"The diagrams brought in by the meter man in the morning indicate to within a few minutes the time the inspectors began and finished the work during the night, the time of closing each stop-cock, and the quantity and kind of the gross and detail waste indicated on the night inspector's report. The night inspection is followed by the day inspection, for which there are also four gangs, who take with them tools to do slight repairs without cost to the tenants. Burst mains, defective valves, hydrants, stand-pipes, etc., are repaired by gangs of pipelayers. In the case of burst or damaged pipes or fittings in houses too serious to be repaired by the day inspectors, notices are issued to repair them in a given time at the cost of the tenant or landlord, subject to the regulations of the water authorities. The day inspector's report is made out on the same book as the night inspector's.

"To insure that these notices are observed two re-examiners are intrusted with the duty of going through the districts which have been inspected, and ascertaining whether the repairs have been properly made. They also examine any new or altered fittings. Some districts may run for a year without inspection, while in others an inspection may be required every month, but no district is allowed to go more than a month or six weeks without the taking of a meter diagram."

An important feature in the Liverpool system is the inspection of fittings. The permanent testing office force consists of seven men—one superintendent, four assistants, and two laborers. The charges for testing are : for bibbs or taps, twopence each ; ball-cocks, threepence ; water-closet cisterns, sixpence. Every tap is taken apart, no matter who makes it, its seat is examined, also the character of the washer, to see that a good bearing is obtained, also the threads on the caps and spindles—indeed, in all respects it is required to conform to the sample tap that the manufacturer had deposited in the office as his standard, which tap had originally received the approval of the engineer. When new fittings are submitted for approval which deviate from the recognized articles, the custom is to have them put in use somewhere and tried for a year, the authorities taking the ground that experiments to demonstrate the value and utility of an article should be made without risk to them or to the house-owner, and though sometimes a meritorious thing is held back a year, yet by this plan a number of unreliable articles are excluded from use, and wrangles with manufacturers and inventors are avoided. In 1880 the number of fittings that " passed through the testing and stamping department was 83,613, of which the testing officers stamped 71,969, the remaining 11,644 being rejected in consequence of defects revealed by the tests applied."

Each approved fitting has two stamps, one that is readily visible and one a private mark. The latter was made necessary because dealers were caught counterfeiting the stamp. In the testing of ball-cocks, they are required to close against a pressure of 130 pounds to the inch, with the ball not over five-eighths submerged; they are also placed upside down, held shut, and subjected to 300 pounds pressure, to ascertain if the casting is perfect. To save the time of immersing each ball, the resistance in weight is estimated and the submerging test is dispensed with.

They have a scale of weight of the water displaced by the ball when immersed five-eighths in depth, a copy of which we were permitted to take by Mr. Davis, the superintendent, which is as follows:

Diameter of Ball.	Weight of Displaced Water.
4½ inches.	18.96 ounces.
5 "	25.89 "
5½ "	34.79 "
5¾ "	39.40 "
6 "	44.76 "
7 "	71.08 "
7⅜ "	83.13 "
8¼ "	115.78 "

To ascertain the necessary weight to put on the ball when it is turned upside down to resist the pressure of 130 pounds, they subtract twice the ascertained weight of the lever and ball from the figures in ounces given on the scale.

Considerable pains are taken in the examination of water-closet cisterns, and in this respect the standard required in Liverpool is higher than anywhere else we know of. A specification was prepared, giving the most minute details for the information of manufacturers, and this must be strictly conformed to, it being held, and very properly, too, that if not accurately made, these appliances are waste-preventers in name only, and on this account we have taken the ground that unless the waste-preventing cisterns contemplated to come within the minimum tax fixed by the Department of Public Works are carefully examined and tested, they will fail to secure the results aimed at ; indeed, for that matter, without careful inspection and good management the whole movement, which is susceptible of doing a great good, will prove disappointing.*

The various patent so-called waste-preventing valves are not allowed. Prior to 1873 single cisterns were used, with one valve, and so arranged that the ball-cock was held up when the valve was lifted. These were so poorly made that they proved unreliable, and they were first opposed because they were provided with a ground-key ball-cock, it not being considered practicable to adjust levers with a ball-cock that came to a dead stop, and as the ground-key ball-cocks were constantly leaking, this fact influenced the prohibition of the single cistern.

In each waste-preventing or double-valve cistern there must be three-sixteenths of an inch play after the valve is closed before it can be approved, and this is carefully looked after by the testing officer. The rules prescribe that the measuring compartment shall hold two gallons, which is ample for the hopper forms of water-closets, but where special kinds of water-closets are used that need a little more, special tanks are sanctioned so long as they are so constructed that there is a limit to the amount that can be drawn at one lifting of the valve. The main point sought after is to prevent a constant running of the water should any one fasten the handle up, a practice often resorted to by persons who have defective plumbing, and imagine that a trickling of water will "keep down the sewer-gas."

It should be remembered that in Liverpool, as elsewhere, much opposition at first prevailed, but the Water Committee, which had the respect and esteem of the community, loyally stood by their engineers and gave the support that a public official everywhere needs when required to enforce what are at the moment unpopular measures, and nowhere is that support more needed than in a city like New York.

It is but justice to state that much of the credit of the success achieved in Liverpool was due to a Mr. Wilson, since deceased, who was indefatigable in his efforts to facilitate the methods of inspection, and to secure the adoption of the details essential to the proper enforcement of the necessary measures.

As in the other cities we have described, no fittings, or class of fittings, are selected that any firm has a monopoly of making—though it often happens that patents are procured on some detail, it is never on

---

* See proposed water-closet tax, New York Department Public Works, Appendix.

any *essential feature required by the authorities*—consequently the widest latitude is given to the purchaser, he being simply required to use articles that are well made, water-tight, and capable of fulfilling the requirements of such devices. Whenever, therefore, the authorities in this country attempt to prescribe the character of fittings, they will find the domain covered by no patents sufficiently wide to secure what the public need, and what any manufacturer can make if he chooses to, at least so far as plumbing fittings are concerned.

In 1881 we learned from Mr. Parry, the efficient engineer then in charge of the distribution of water, that 122,673 houses were supplied with water by the corporation; the population was 725,000, being an average of 5.91 persons to each house. The average rate of consumption for all purposes was 23 imperial gallons per head daily, which included five gallons per head for trade purposes measured by meter, about three gallons for sanitary purposes and miscellaneous trade uses, charged for by assessment, leaving about fifteen gallons per head daily for domestic purposes, stores, hotels, public buildings, and all waste. Water-closets are general in Liverpool, and each one is required to be so fitted that it can be flushed with two gallons in a few seconds; private baths are common, and of late years it has been the practice to put them in all new houses renting for more than $90 per year. There is a considerable demand for shipping, and public fountains are numerous. In order to sum up the results secured by the adoption of efficient measures to restrict waste we cannot do better than quote from Mr. Thomas Hawksley's address as President of the National Association for the Promotion of Social Science, delivered at Liverpool, in October, 1876:

"Under the able management of Mr. Deacon (Borough Engineer) the supply has been changed with great advantage from a restricted intermittent to an unrestricted constant system, and as a consequence of the admirable method he has devised and adopted for the discovery, suppression, and prevention of waste, the municipality has emerged, as respects the quantity of water at its disposal, from a state of actual poverty—poverty in the midst of plenty—to a condition of redundant wealth. It has enough and to spare."

"To provide a town with water of a suitable quality is now a matter of great difficulty and much expense. Competent and available sources are becoming fewer and fewer, and must in general be sought at distances always becoming more remote. Moreover, the value of land, labor, and materials is almost day by day advancing, and so rapidly and steadily, that the present cost of a reservoir is, as I know by repeated experiences, double what it was less than thirty years ago."

In view of these facts so forcibly stated, is it not stupid to clamor that "water should be free as air," and should it not be considered a crime to waste and deprive others of what they are entitled to?

CHAPTER V.

PROVIDENCE AND CINCINNATI.

REVIEW OF MEASURES TO PREVENT WATER WASTE IN THE UNITED STATES PRIOR TO 1882—PROVIDENCE, R. I., RESULTS FOLLOWING THE GENERAL USE OF METERS—CINCINNATI, METHODS OF HOUSE INSPECTION WITH THE AID OF WATER-PHONE—RESULTS OBTAINED.

HAVING in preceding chapters pointed out what has been accomplished in the way of preventing waste of water by the adoption and enforcement of efficient measures in Great Britain, we will now briefly review the situation in this country. In looking over the field, we are bound to admit that, with the exception of two cities, hardly a beginning has been made, though considerable has been said on the subject in print, and a scarcity of supply has forced the consideration of the matter on engineers.

Providence shows the best record of any city in this country so far as the amount of water consumed goes; though it is true the introduction of public water is of comparatively recent date (1870), and from the commencement meters have been in general use. The amount used in 1883 was 40 gallons per head, or 379 gallons per tap daily, while in Milwaukee, a city of about ten thousand more population, with no special regulations to restrict waste, the consumption amounted to 128 gallons per head, or 1,476 gallons per tap daily, Milwaukee having 350 meters, while Providence had 5,721.* Such a showing indicates what it is possible to secure by good management.

The table on page 26, for which we are indebted to Mr. Samuel A. Gray, City Engineer, was prepared in response to certain inquiries made by us, and will be found of special interest. It certainly makes a good showing for the meter system.

We believe that the scale of water-rates is so arranged that most consumers elect to have meters. They find that with care their tax is

---
* From "*Statistical Tables of American Water-Works,*" by J. J. R. Croes, C. E.

CITY ENGINEER'S OFFICE, WATER DEPARTMENT, PROVIDENCE, R. I., MARCH 7, 1883.

Data relating to Meters, Daily Consumption of Water, etc., for the past Six Years. Providence Water-Works.

Year.	Estimated Population to Middle of the Year.	Number of Meters in Use December 31.	Net Amount Charged to Meter Department for Setting and Repairing Meters, etc.*	Number of Services Opened to December 31.†	Total Water Receipts for the Year.	Average Daily Consumption of Water During the Year.	Gallons per Capita.	Per Cent. of Meters to Services Opened.	Estimated per Cent. of Length of Received and Platted Streets and the Received Streets to the Length of Distribution Pipe Laid.		Remarks.
									Received and Platted.	Received.	
1877	101,000	3,203	$476 40	7,789	$200,039 39	2,492,032	24.7	41.1	49	93	
1878	102,286	3,648	602 34	8,566	218,883 33	2,701,404	26.4	42.6	50	95	
1879	103,572	4,036	805 37	9,139	229,551 78	3,110,279	30.0	44.2	49	96	
1880	104,857	4,452	324 35	9,757	247,705 06	3,547,264	33.8	45.6	49	97	Water-rates reduced.
1881	109,671	4,784	111 99	10,305	260,530 87	3,716,937	33.9	46.4	51	98	
1882	114,377	5,279	57 39	10,919	269,318 77	3,665,427	32.0	48.3	51	100	Water-rates reduced.
			$2,377 84								

* Yearly balance from the books of the Water Board. For individual years it is only approximate, as the amounts are not entered until the bills are paid, which is more or less in the following year. The total sum of the six years is very nearly correct, but from this amount also should be deducted some charges which were paid early in 1883. The employees of the Meter Department are at times called upon to do other work than that actually pertaining to meters, but they do not read the meters.

† The actual number of services in use December 31, 1882, was 10,357, and the difference between 10,357 and 10,919 should be the number closed, disconnected, etc., since the first service was opened in 1871; therefore, the actual per cent. of services now open, metered to December 31, 1883, is 51, instead of 48.3 per cent, as above.

All of the above figures include data for 76 miles of pipe supplied with water lying outside of the city limits, with the exception of population and percentage of streets in which pipe is laid, which include only the city proper.

less than if they pay for each fitting in a building in accordance with the published schedule.

Cincinnati is another place that has accomplished something in reducing the amount of waste, and that seems to be mainly due to the efforts of the authorities, aided by the use of the Bell waterphone, to discover where abuses existed and waste was occurring.

During the recent flood in that city, it will be remembered, the pumping-engines were stopped by the high water, and a very possible calamity averted as the direct results of their waste-prevention efforts. It seems that when the pumps stopped there were about one hundred million gallons in the reservoir, and when pumping was resumed there were forty millions left, and the engines were stopped five days. Had nothing been done to check waste, it will be seen that even this very slender reserve could not have been had, for taking the ordinary increase on the basis of former years' consumption, the original one hundred millions would have been exhausted before the engines were started. After this experience, we imagine the people of Cincinnati will more thoroughly appreciate the value of vigorous waste-prevention measures.

In this connection the following data and description of the methods employed may be of interest.

They began the waste inspection with the aid of waterphones the last of November, 1881, although it was not systematically done until July, 1882. The population at that time was estimated at 264,000; it is now estimated at 280,000.

The daily consumption for November, 1881, was 19,706,903 gallons, that being the average for the month. The daily average for January, 1883, was 16,264,000 gallons.

There are employed six men specially on waste inspection and three others at irregular intervals. The annual cost is $5,500.

The method of operation is as follows:

The inspector, with assistant, is furnished a record book, containing the number of house, service-stop and location of same, and blank columns for record of inspection, a hydrant-key, lantern, and waterphone. He begins work at 11 o'clock each night and stops at 4:30 A. M. He reports at office at 8 o'clock the results of night inspection, and then views, or rather inspects, the various premises where running water was detected the previous night. His record of the night work shows him the character of the flow—viz. :

First—Whether it is a large flow or abuse, or a small one or leak.

Second—Whether it is in street or outside of stop, or inside of premises.

He first examines for abuses, then visible leaks, after which he tests for underground leaks, and then makes a record accordingly, opposite the premises. If there is an abuse he leaves a notice to that effect ; if a leak he gives a leak notice.

We are indebted to the officials for the following data, which shows the comparative consumption of water (daily average) since 1879 :

	JANUARY.	FEBRUARY.
	Gallons.	Gallons.
1879	15,430,638	15,198,733
1880	15,569,532	15,452,402
1881	18,406,655	19,515,450
1882	17,287,479	17,751,835
1883	16,264,000	15,320,687

The authorities do not presume to regulate, inspect, or interfere with the plumbing inside of houses, further than to require defective fittings repaired or replaced. Plumbers are licensed and required to use the regulation service-pipes, sidewalk stop and box ; otherwise no discrimination is made, except as to hopper-closets, which are charged double the usual rates.

If a fitting is found a second time out of order, or, in the judgment of the inspector, it cannot be repaired, it is condemned.

The following was the consumption of water for 1882 :

	1881.	1882.
	Gallons.	Gallons.
Total pumpage	8,623,788,320	7,232,134,700
Deduct injection-water for low-pressure engines.	190,994,120	99,215,440
Deduct differences in coils of reservoir	35,140,000	6,350,000
Total actual consumption of water	8,397,654,200	7,126,569,260

Daily Averages.

	1881.	1882.
	Gallons.	Gallons.
January	18,406,655	17,287,479
February	19,515,450	17,551,835
March	18,892,087	17,493,951
April	19,283,414	19,801,381
May	23,534,065	19,804,432
June	23,484,225	20,685,542
July	28,507,603	21,255,538
August	31,665,708	22,246,513
September	28,871,689	22,004,314
October	25,802,557	21,374,545
November	19,706,973	18,005,442
December	18,056,895	16,809,083
	23,007,272	19,524,848

Consumption per Capita.

	1879.	1880.	1881.	1882.
For general surveyed rents	26.5	27.0	27.7	28.0
For metered rents	5.0	6.7	6.0	8.5
For hydraulic-elevator rents	3.0	4.9	4.0	5.0
For street-sprinkling rents	0.7	0.8	0.6	0.7
For free water, fire purposes, etc.	2.0	2.7	3.1	3.2
For waste	30.8	32.9	45.6	24.3
Total	68.0	75.0	87.0	69.7

Classification of abuses and leaks discovered and corrected for the year 1882:

Willful cases, abuses in water-closets	170	
Willful cases, abuses in hydrants	211	
Willful cases, abuses in urinals	19	
		400
Leaks in water-closets	389	
Leaks in faucets	771	
Leaks in hydrants	1,107	
Leaks in hose-plugs	119	
Leaks in lead pipes	286	
		2,672
Underground leaks in lead pipes	299	
Underground leaks in stops	86	
Underground leaks in hydrant-stocks	1,035	
		1,420
Miscellaneous leaks		314
Total abuses and leaks		4,806

[Several other cities have a large proportion of their connections metered. In Worcester, Mass., 80 per cent., in Fall River, Mass., 66 per cent., and in Pawtucket, R. I., 60 per cent. of the services are supplied through meters. In Taunton, Mass., 25 per cent., and in Yonkers, N. Y., 58 per cent. of the services are metered, with very good results.* May, 1885.]

---

* See "*Statistical Tables of American Water-Works,*" by *J. J. R. Croes, C. E.*

CHAPTER VI.

NEW YORK.

MEASURES ADOPTED BY THE DEPARTMENT OF PUBLIC WORKS PRIOR TO 1882.

The following letter from the Commissioner of Public Works of New York City explains what the Department has been doing in the matter of restricting waste :

> DEPARTMENT OF PUBLIC WORKS,
> COMMISSIONER'S OFFICE, No. 31 CHAMBERS STREET,
> NEW YORK, February 21, 1883.

To the Editor of THE SANITARY ENGINEER :

' SIR : In answer to your inquiries in reference to the subject of waste water in this city, I beg to say :

The first practical steps against waste of water were taken in 1876, by establishing a system of house inspection to detect and prevent leaks in plumbing and willful waste in letting water run from faucets unnecessarily.

The next step was the abrogation, in 1877, of the contract or license system of supplying water to shipping and for building purposes, by which immense quantities of water were wasted along the water-front and in the erection of buildings. This service is now under the immediate supervision of competent inspectors, and the greater portion of the water supplied to shipping is measured by meters.

The last and most effective measure against waste is the use of water-meters. On January 1, 1877, there were only 260 water-meters in use in the city. In 1877 and 1878 the number was increased to 922; in 1879 to 1,398 ; in 1880 to 4,002 ; in 1881 to 5,293 ; and at this date there are 6,924 meters in use.

The expenditures of the various measures for the suppression of waste are paid from the appropriation for " Repairs and Renewals of Pipes, Stop-cocks, etc.," which is the appropriation for the repairs and maintenance of the entire distributing system, comprising 531 miles of water-pipes, with 5,613 stop-cocks, and 6,944 fire-hydrants.

The Department estimate for this appropriation for the year 1883 was as follows:

Four repair companies, each consisting of one foreman, two carts, and twenty-one mechanics and laborers......	$82,000
One gang of mechanics and laborers, employed in placing taps in water-pipes, and cutting off taps for non-payment of water-rents, waste of water, etc............	24,000
For new 12-inch and 6-inch pipes, stop-cocks, hydrants, fixtures, materials, and supplies for renewals and repairs..	122,000
For services of engineers, meter inspectors, inspectors of waste, superintendent, and inspectors on water supply to shipping, clerks on water-meter accounts, etc..	72,000
Total.................................	$300,000

The Board of Estimate and Apportionment in the final estimate appropriated only $170,000.

It is evident that with the constant and rapid growth of the city, the increased demands on the water service, the addition of from fifteen to twenty miles of pipe each year to the distributing system, and the wear of the older pipes and stop-cocks, the expense of repairs and maintenance must increase from year to year, and the necessity for further suppression of waste, in order to meet the new demands on the water supply, grows in the same proportion.

For years past the appropriations have been inadequate to afford means for the renewal of the old pipes and stop-cocks along the water-fronts, which are so reduced by corrosion from contact with salt water that they are unsafe.

The measures for the suppression of waste began with the places and establishments where the consumption and waste of water were greatest—the docks, gas-works, breweries, malt-houses, railroads, hotels, factories, large stables, etc. As the work is extended to the smaller and more numerous business places, it becomes more laborious in proportion to the results which can and which must be attained to keep up an efficient supply.

It will be impossible, with the amount appropriated by the Board of Estimate and Apportionment, to perform the various work and duties referred to, to the extent and with that degree of promptitude and efficiency which the situation calls for.

In carrying out the various measures for the suppression of waste of water the department now employs—

On water meters:
One general inspector to examine places after they are metered.
Twelve inspectors to read meters and keep them in order.
One inspector to test meters in pipe-yard.
Three clerks on meter accounts.

On waste of water:
One general inspector.
Eighteen inspectors to examine plumbing in buildings.
Four inspectors to examine house-drains in sewers at night.

One inspector to examine new buildings.
One inspector to visit houses where water is found wasting through drains at night.

On supply of water to shipping, etc.:
One superintendent.
One clerk keeping accounts.
Seven inspectors along the water-front and at new buildings.

Very respectfully,
HUBERT O. THOMPSON,
Commissioner of Public Works.

The foregoing omits to state one method adopted to check waste, which, after all, is the principal one, aside from the use of the 6,924 meters, that has secured any practical results—that is, the throttling of the distributing-pipe or partial closing of the gates from the Central Park reservoir. The engineers, some time since, realized that unless they kept the pressure down in the mains, except in case of a fire, the water would be drawn out of the reservoir faster than it could be delivered to it—in other words, 95,000,000 gallons was all that could run into it in twenty-four hours. It was therefore necessary to so throttle the distributing-pipes as to limit the quantity withdrawn to that amount, or else the city would be without any reserve whatever, a contingency not pleasant to contemplate, with the possibility of breaks in the aqueduct. It may be asked, Why throttle the pipes and not give us what pressure there is? We would reply that, with the present worn-out plumbing and leaking fittings all over this city, and defective pipes under the streets, the leakage would increase as the pressure was increased, and, in our opinion, if a pressure of forty pounds per square inch were maintained throughout the greater part of the city, assuming such a thing possible, the amount of water consumed would be at least 125,000,000 gallons per day, instead of 95,000,000 as it is now. It is, therefore, evident that we must do as was done in Norwich, Liverpool, and other English cities, and that is, stop the leaks in the houses and underground, and keep them stopped, or any additional pressure, when we do get it, will simply increase the waste, and so much of it as soaks in the ground will moisten the soil about our habitations and increase the mortality from lung diseases.

[There is reason to believe that considerable work has been accomplished within the past two years. The number of meters has largely increased, there being over 11,000 now in use, or about 12 per cent. of the total number of house-connections. May, 1885.]

CHAPTER VII.

GENERAL CONCLUSIONS.

In the preceding chapters we have endeavored to show what, in the absence of stringent regulations, is inevitable in any community that has a public water supply—namely, that habits of wastefulness are formed, grow, and become difficult to control; that sooner or later, as a direct consequence, scarcity prevails and demands exceed anticipation. Hitherto these demands have been met by taking possession of available sources of supply or increasing pumping facilities. Such sources, however, are rapidly becoming scarce, and the pumping is found to be expensive; indeed, our conditions are rapidly assuming the proportions found in Great Britain, and, like Great Britain, we must grapple with this problem of waste and its prevention.

The experience obtained there we have briefly outlined in the foregoing chapters, and, so far as we are acquainted with the facts, we have given our American cities credit for what they have done in that direction, though we regret so little is to be recorded. It now remains to enumerate the methods that might be made available in the city of New York—provided the needed money and powers be granted—and to suggest the various conditions involved that seem to require consideration.

First—The placing of a meter in every house, assuming no control over plumbing, but relying on consumers' desire to economize in water-tax, to secure a stoppage of waste, to cause the reconstruction of defective plumbing and the removal of wasteful fixtures.

This plan suggests the following points for consideration :

Will the knowledge that water is to be paid for by measurement have the effect of preventing any considerable number of our people from bathing and from using sufficient to properly flush water-closets and urinals, thus producing insanitary conditions?

Can this be obviated by allowing an ample amount per capita for legitimate use, for which a specified tax is to be made, said tax to be paid whether this water is used or not, but whatever may be used in addition thereto to be paid for by measurement?

Since large numbers of consumers are tenants, moving every year, are they likely to find it cheaper to retain leaking fittings, wasteful fixtures, and pipes exposed to frost, paying the water-rates required, rather than go to the expense of reconstructing the plumbing in a house owned by another in which they have but a temporary interest?

Would a consideration of the foregoing indicate that the use of meters would make the supervision of plumbing in buildings superfluous?

What measures should be adopted to ascertain when waste was taking place underground?

What would be the probable expense of placing meters in the 95,000 houses in New York—that is, cost of meters and expense of setting them—and what time would be required to do it?

What would be a fair amount to charge for interest on the outlay, cost of repairs and maintenance, and what amount per year should be allowed for wear and tear?

What amount would be required for the clerical staff to keep the accounts and render quarterly bills, which would seem necessary if the bills are to be relied on to enforce care in the use of water?

What number of inspectors would be required to take the readings of the meters?

Should the city pay for meters, if it is decided that they be used, or should the Providence plan be resorted to—namely, revise the water-rates so that a charge be made for every tap and fixture, unless the house-owner elects to accept and pay for a meter, it being understood that a specified tax shall be paid whether water is used or not, but only the excess of this specified amount to be charged for by measurement?

Second—If the general use of meters in this city is objected to, and systems of inspection are to be relied on, are the inspections to be made only where waste is indicated, by the adoption of some agency like the Deacon meter, Bell's waterphone, Church's indicator, or any other device, to indicate the flow of liquids?

Are examinations of premises to be restricted to those in which some outside agency indicates waste is occurring?

Would the interest on the cost of adopting and maintaining some outside agency to determine where waste, either underground or in a building, was prevailing, be more than offset by the facility of reaching leaks, the probable employment of a smaller force of inspectors, the being able to have a check on their operations, and the exemption of householders whose fittings are in good order from periodical visits and the annoyance attendant thereon?

If fittings were required to conform to a standard, to be examined and stamped (fees to be charged therefor), would not the occupants of the houses, or those who buy them, be saved the expense of frequent repairs, early renewals, and would waste in any measure be prevented?

Is it possible in a large city, with a large moving population, occupying buildings erected or owned by others, to secure the proper location of pipes and the use of proper fittings in the absence of rules rigidly and constantly enforced?

With the conditions existing in New York, what outdoor agency to indicate when waste was taking place, either in or outside buildings, could be adopted with the best advantage, taking into account cost and results to be obtained?

Third—If the former two plans are deemed impracticable, is there any that can be adopted that will secure any reasonable results other than the Manchester system, which requires the testing and stamping of fittings, that plumbing-work be done in accordance with prescribed rules, and that all wasting or wasteful fittings be replaced by approved and proper ones, and that all buildings and all parts of buildings where water flows shall be frequently inspected? And must the annoyance to house-owners therein involved be imposed?

Would any system secure adequate results that did not involve the testing and stamping of fittings, supervision and control of plumbing, inspection of buildings, repair or removal of defective and wasteful fixtures?

Which of the plans described would involve the constant employment of the smallest permanent staff in proportion to the results obtainable?

We have here indicated some of the questions to be considered by those who may be required to adopt and sanction a plan for cities like

New York that require consideration. It will be evident that, no matter which method is adopted, a considerable expenditure of money is inevitable, and that with one scheme the first outlay may be larger and the cost of maintenance more, while another, though probably more annoying to householders, involves a smaller investment and a less cost of maintenance, though to obtain adequate results, it requires a better system of admistration and a better class of men than have hitherto been selected for such duties.

It is clear that something must be done, and the longer it is put off the more complicated becomes the problem and the greater the final expense.

A demand from the Commissioner of Public Works for the needed money to carry out any effective plan will, in the present condition of public sentiment, meet with opposition and suspicion.*

Without discussing the causes that produce this feeling, we are of the opinion that the best thing to be done under the circumstances is to have the Mayor appoint a commission of three gentlemen, who have the confidence of the public and the profession, one of whom, at least, shall be a civil engineer of experience in the matter of water supply and distribution, and who is not directly or indirectly interested or identified with any system or plan. These gentlemen should take testimony and investigate the subject thoroughly, and, as soon as practicable, recommend a plan. They should also submit a bill for enactment, if any additional legislation seems to them necessary.

If this is done, and any reasonable plan honestly and efficiently carried out and enforced, this city will be relieved from severe burdens and very possible calamities ; and if the information we have gathered and recorded will in the slightest degree contribute anything toward the accomplishment of so desirable a result, we shall feel amply repaid for the labor involved in the preparation of these articles.

---

* This remark, written in 1883, is equally true at this time, and has been abundantly proven by the comments of the daily press and the action of the Legislature on the request made a few days ago by the present Commissioner. May, 1885.

APPENDIX.

WATER-WASTE PREVENTION IN BOSTON.

FOLLOWING the publication of the foregoing articles a new water board was appointed in Boston, which, having the records and reports of the city engineers for several years prior to stimulate its action, and the evidence to demonstrate what had been done elsewhere, secured from the City Council power and funds, which it has utilized in a very satisfactory manner, as is shown by the report of its first year's operations, which was published in THE SANITARY ENGINEER, of November 13, 1884.

CHECKING WASTE OF WATER IN BOSTON.*

THE present board early realized that the prevention of waste was one of the most important and difficult problems with which all large water-consuming communities are obliged to contend. It is conceded by all experienced observers that at least 40 per cent. of the water supplied to large towns and cities is willfully wasted. The majority of people seem to be possessed of the idea that water should be supplied as free as air, and hence all idea of economy in its use seems to be banished. Restrictive measures have seldom been applied in our country, and the result is that everybody has become extravagant and wasteful.

We stated in our report of September last that the enormous wastage constantly taking place had been brought to the attention of the City Council every year since 1852, but that practically nothing had been done to remedy it. The Joint Standing Committee on Water, in its report of April 30, 1883, urged that immediate measures should be taken to stop this waste and reduce consumption. We believed that the consumption should be reduced from 95 to at least 60 gallons per

---

* From the eighth annual report of the Boston Water Board for the year ending April 30, 1884, but including the operations for checking waste up to September 1, 1884.

capita, and that if this could be done a very large amount of money would be saved to the taxpayers of the city. We accordingly organized early in July, 1883, the Division of Inspection and Waste. The work accomplished by this division last year was of very great importance to the city; indeed, without it we should have been compelled to cut off the supply during a part of the severe drought of last year from a considerable number of our citizens. The tabulated results of the labor of this division appear in our reports of September and December last, and in the appended report of Superintendent Cashman. This report confirms the correctness of the judgment and policy of the board, and shows that continuous systematic inspection is an essential element in the prevention of the wanton waste now so prevalent in all large communities. Under this inspection the premises and fixtures of every water-taker have been visited several times during the year, the leakages stopped, the defective pipes and apparatus repaired, and the people taught to respect and obey the city ordinances with reference to the prevention of waste.

During a part of the present year, since the date of Mr. Cashman's report, the Deacon system of waste-detection has been in operation in conjunction with the house-to-house inspection, and up to the present writing, September 1, the results have been very satisfactory.

SUDBURY AND COCHITUATE WORKS.

	1882.		1883.		1884.	
	Daily average consumption.	Gallons per head per day.	Daily average consumption.	Gallons per head per day.	Daily average consumption.	Gallons per head per day.
January	32,151,100	92.9	34,715,500	97.8	32,162,300	88.4
February	34,662,300	102.2	32,690,700	92.0	24,598,000	67.5
March	32,656,300	94.1	34,110,700	95.8	23,711,900	65.0
April	30,827,000	88.6	30,617,600	85.8	21,505,700	58.8
May	28,738,000	82.3	32,169,500	89.8	23,708,500	64.6
June	33,178,400	94.8	33,419,200	93.3	26,184,600	71.2
July	30,992,600	88.5	36,774,000	102.4	25,409,000	68.9
August	34,149,300	97.3	37,141,000	103.2	25,065,200	67.7
September	31,691,600	90.0	33,645,600	93.2	....	....
October	31,563,800	89.4	29,575,800	81.9	....	....
November	31,138,700	88.7	28,839,300	79.6	....	....
December	32,352,300	91.4	30,174,200	83.0	....	....

Mystic Works.

	1883.		1884.	
	Daily average consumption.	Gallons per head per day.	Daily average consumption.	Gallons per head per day.
January............................	8,369,600	97.3	8,019,100	92.2
February...........................	7,714,650	89.6	6,349,500	72.9
March..............................	7,737,300	89.8	6,337,100	72.7
April...............................	6,171,150	71.5	5,242,100	60.1
May................................	6,319,100	73.1	5,800,000	66.4
June...............................	6,912,550	80.0	6,245,600	71.5
July................................	7,307,550	84.5	6,312,300	72.1
August.............................	7,261,500	83.9	6,088,400	69.5
September.........................	5,846,300	67.4	....	....
October............................	5,497,250	63.4	....	....
November..........................	5,930,600	68.3	....	....
December..........................	6,771,500	77.9	....	....

The tables show the daily aggregate and per capita consumption in the Cochituate and Mystic departments for the first eight months of 1884 in comparison with the corresponding months of 1883, this period being covered by the present system of inspection; and also a statement of the average daily consumption for several months prior to the beginning of the work of inspection.

The tables show the daily average consumption in the Cochituate supply from January 1, 1883, to September, 1883, to have been.. 33,954,775 gals.
And for the corresponding period of the present year, under the inspection system....................... 25,293,150 "
                                                                                                          "
A net daily average reduction of.................. 8,661,625 "
or about 26 per cent. saving.
In the Mystic Department, for the same period in 1883, the average daily consumption was.... 7,224,175 ·
And for the corresponding period in 1884, 6,299,262

A net daily average reduction of 13 per cent., or... 924,913 "
                                                                                                          "
Making a total average daily reduction............. 9,586,538 "

It is computed that the actual cost to the city for each 100 gallons of water furnished is about $1\frac{26}{100}$ cents, and upon this basis the reduction in consumption represents in round numbers the sum of $1,200 per day during the first eight months of 1884. Again, it will be observed that in the months of July and August, which may properly be cited at this writing, the Deacon system being in full operation, the consumption was reduced 34 and 36 gallons per head each day for those months respectively.

It may be interesting to make a single comparison between the consumption of 1882 and 1883, when no special efforts were in progress to economize the supply.

The daily average consumption in the Cochituate supply from January 1, 1882, to September of the same year, was.......... 32,116,288 gals.
And for the corresponding period in 1883 it was.... 33,920,422 "

Making an increased daily average consumption of.. 1,804,134 "
In the Mystic supply the daily average consumption for the year 1882 was............................. 6,574,400 "
The daily average for the first eight months of 1883 (being the non-inspection period) was.... .... 7,224,175 "

A net daily average increase of................... 649,775 "
And the net daily average increase of 1882 over 1881 was... 330,300 "
The daily average consumption in the Mystic for the first eight months of 1883 (the non-inspection period) was.. 7,224,175 "
And for the last four months after inspection began.. 6,011,412 "

A daily average reduction of...................... 1,212,763 "
The average consumption per head per day for the year 1882 (discarding fractions) was................ 89 "
And for the non-inspection period of 1883, eight months, from January to September, it was........... 92¼ "
For the corresponding period of 1884... .... 69 "

Showing a net average reduction of 20 gallons per head in 1884 over 1882, and 23½ gallons over 1883.

The large decrease in consumption materially lessens the cost of pumping, the saving in coal at the Highland pumping-station alone being 35 per cent. for the first eight months of 1884 over the corresponding period of last year.

These results demonstrate beyond question the wisdom, as well as the necessity, for the adoption of measures to check the prodigal waste which prevails. Unless some radical system be adopted, which will keep the consumption down to 60 gallons or less per capita, the taxpayers of Boston will be compelled, at a very early day, to expend several hundred thousand dollars in the erection of additional reservoirs, and several millions of dollars in obtaining a new source of supply. The city of Providence, with relatively the same industrial interests and class of people as our own community, keeps her consumption down to about

36 gallons per head per day. That of Boston for July, 1883, was 102 gallons per head each day, and in July of the present year, under the inspection system, 69 gallons per head. It will be observed that, even with this great reduction of the present year, we are still using nearly 100 per cent. greater amount of water per capita than our more prudent neighbor. In Providence, however, more than 50 per cent. of her consumers are supplied by the meter or measurement system, while with us scarcely 10 per cent. are supplied by this method. Providence is, perhaps, the only city in the United States which has undertaken to manage its water interests with the same reference to business principles that prevails among business men everywhere.

In connection with the house-to-house inspection system, and as a necessary supplement thereto, the board adopted the policy recommended in its December report, as follows :

First—To put recording meters on all manufactories, breweries, stores, business establishments, hotels, tenement-houses, and all other places where a large quantity of water is used, or where waste prevails.

Second—To establish Deacon waste-detectors in the residential portions of the territory supplied with water, making specific districts, and doing the work in conjunction with the house-to-house inspection.

Third—To begin the putting in of sidewalk stop-cocks at once, adopting the Church stop-cock as the best complement to the Deacon, if the further trials continue to prove its excellence.

Recording-meters have been applied in accordance with this original plan. The meters used have been largely those of the Tremont pattern. They have been delivered as fast as the city could use them, and have proven generally satisfactory both with reference to workmanship and accuracy.

The Deacon system of waste-detectors is at this writing (September 1) fully applied to the Cochituate Department. This system was thoroughly tested in the Mystic Department in 1882 by Assistant Engineer Dexter Brackett, and was recommended by Mr. Henry M. Wightman, the engineer of the board, and Alderman Greenough, the Chairman of the Water Committee, before the present Water Board was appointed. To these gentlemen should be given in large part the credit of introducing this very efficient system into Boston. We now have some seventy-five of the Deacon detectors in operation, and the work being done by them is of a satisfactory character.

The board was directed by the City Council, on the 19th of December, 1882, to make an examination of the merits of the Church stopcock, and a very thorough series of tests were made last year under the supervision of Assistant Engineer Dexter Brackett, and a full report of the same appeared in the report of December 6, 1883, page 40. Since that time Mr. Henry M. Wightman, engineer of the board, has made various experiments with this stop-cock, and several important changes have been made in it as the result of his suggestions. The inventor, Chief Engineer B. S. Church, of the New York Aqueduct Commission, has also made a very valuable addition to the instrument, by means of which the particular floor upon which the water may be leaking or running is indicated on the dial attached to the stop-cock in the sidewalk. The board has been conducting tests and experiments with this invention for more than twelve months, and is satisfied that it is the best stop-cock for general purposes and waste-water detection combined. No city or town having a water-supply should be without a complete system of sidewalk stop-cocks. Every service applied in Boston for the past few years has been accompanied by such a stop-cock. The failure of our water authorities to apply them originally was a grave mistake, and one which no other large community, save New York, has committed. The plan of the board is to gradually apply these instruments until the city has its full complement. At this writing the Church stop-cock has been adopted by the board, and 5,000 of them have been ordered. The necessity of the early application of sidewalk stop-cocks was presented in our report of December (page 48), and has frequently been urged by Engineer Wightman.

We have also made extensive experiments with the Bell waterphone, which has been used very successfully in Cincinnati and Philadelphia, and is about to be introduced in New York. Our experiments, however, were not successful, for the reason that, in the absence of sidewalk stop-cocks, wires were used to connect the waterphone with the service-pipe, and the results were unsatisfactory. We contemplate giving this system another trial at an early day, in the Mystic Department, where sidewalk stop-cocks exist, and where the same conditions will obtain as in other cities in which it has met with success.

It is a matter of frequent complaint that the system of house-to-house inspection established by the board is annoying to water-takers; the frequent visitations of inspectors to premises of water-takers are looked upon as a system of espionage. There may be some measure of truth in these complaints, but there is no present remedy. There is but one other large city in the country so deficient in sidewalk stop-cocks as Boston; and until these are generally applied and the recording-meter service largely extended, we must continue the house-to-house inspection system, or submit to the prodigal waste which has heretofore prevailed. There is no middle ground. People *will not* repair their defective fixtures, and *will not* stop wasting water unless compelled to do so by official visitation, or by the adoption of a measurement system which will oblige them to pay for all the water used.

Our inspection corps is composed of gentlemanly officials. Each officer is provided with a badge, which must always be worn in sight when on duty, and a commission which he must exhibit on demand. We have frequently had the whole corps of inspectors before the board, specially to instruct them with reference to these matters. Whenever complaints are made in this direction the board causes them to be thoroughly investigated, and prompt action is always taken. The board has used every precaution to insure courteous treatment to the water-takers and secure efficient results to the city. This system, vigorously followed up, will prevent a large part of the daily waste; and no other method will accomplish it except the recording-meter or measurement system. When this is applied, and people are required to pay for water as they do for gas, they will not waste it. The meter or measurement system could not be universally applied to Boston except at a very large expense, and even then it would require several years.

Recognizing these difficulties the board adopted the only feasible plan of reducing the consumption—that of an efficient house-to-house inspection—and the results have justified its action.

THE SANITARY ENGINEER commented editorially in its issue of November 13, 1884, on the results of this action of the Boston Water Board in the following terms:

"It is gratifying to see that our labors in behalf of water-waste prevention are beginning to show some results in this country. In Volume VII. we published a series of articles showing what had been done in English cities, and in Cincinnati and Providence, by honest, systematic effort. It is, therefore, a pleasure to be able to offer further American evidence in the same direction, and we invite a perusal of an extract from the report of the Boston Water Board, elsewhere printed. We hope that under New York's new municipal management an honest effort will be make to stop the criminal waste now prevailing.

"The report of the Boston Water Board, which is published under date of May 1, 1884, but which recites the operations of the department up to September 1, 1884, in the matter of waste-prevention, is startling as exhibiting the utter falsity of the complaints that have been made for years by managers of public water-supplies in America, that it is impossible to keep consumers from wasting water. They have said so in Boston year after year, just as they have in New York and everywhere else, but the fact is that they have never tried to stop waste except by exhortation, which has proved as effective as a judge's charge to a grand jury is in preventing crime. But the present Boston Water Board has gone to work systematically and sensibly, and has already produced results even more extraordinary than those brought about in Cincinnati by a similar procedure. In Cincinnati the result of a systematic inspection of waste by means of the Bell waterphone, and the subsequent repairing of fixtures leaking by accident or on purpose, reduced the daily consumption 16 per cent. (from 23 million gallons in 1881 to 19.3 million gallons in 1883), and the cost of fuel for pumping from $75,257.63 to $54,671.75. In Boston the application of the Deacon waste-detector on the mains and the Church stop-cock on service-pipes reduced the daily consumption of May, June, July, and August 28 per cent. (from 32.3 million gallons in 1883 to 25.1 million gallons in 1884) in the Cochituate Department, and 12 per cent. in the Mystic Department.

"It is estimated that this reduction of consumption is equivalent to a saving of $1,200 per day during the first eight months of 1884. This saving does not bring inconvenience to a single consumer. The quantity of water supplied for actual use is not curtailed in the slightest degree, and no individual suffers in consequence of the restrictions placed on the immoderate waste of careless water-takers, for which, under the absurd existing system of selling water by guess-work, the careful consumer has to pay more than his proper share. It is time that the officials of other cities where water is wasted either take up the matter of waste-repression in earnest, or are themselves looked after by the taxpayers, who have to foot the bills for needless extensions of works of supply."

WASTE-PREVENTION IN NEW YORK CITY.

(From the report of the Commissioner of Public Works, December 31, 1884.)

CONSUMPTION AND WASTE OF WATER.

"The measures adopted by the Department to detect and check waste of water were pursued to the fullest extent of which the means appropriated for the purpose would admit.

"The Waste Inspectors made 192,277 house inspections, detecting 5,944 places where water was wasted in consequence of defective plumbing, and 272 places where it was wasted by keeping faucets open. They also made 9,275 night examinations of house-drains to find places where water is allowed to run at night; 940 such places were found, the flow of water through the house-drains into the sewers being at the rate of from one to five gallons per minute. In any one of these cases more water is wasted than would supply the legitimate needs of a dozen large families.

"While the measures and efforts of the Department for several years past have undoubtedly resulted in great improvement in respect to the general waste of water, there is one old habit which it seems impossible to check with the present means and authority—viz., that of letting the water run from faucets day and night in cold weather to prevent freezing in the pipes. The enormous extent of this waste is shown by the fact that on one single cold day, December 20, the water in the Central Park reservoir was drawn down five inches, showing a waste of 13,000,000 gallons on that day over the usual consumption and over the supply received through the aqueduct.

"Additional water-meters, the most effective instrument for stopping waste, were placed as rapidly as practicable during the year. The number placed is 2,613, making 11,625 meters in use on December 31."

In reply to objections made by some persons to a house-to-house inspection, THE SANITARY ENGINEER on April 23, 1885, thus commented editorially on the above report :

"But the ground is now taken that it is wrong to check waste because the wasted water cleans the drains and sewers. This is nonsense. The continual dribble of a faucet does not clean out the drains. The emptying of a bath, a basin, or a water-closet does create a flush which will remove obstacles, but the passage of ten times that amount of water distributed over a long period does not have that effect at all.

"In 1884, out of 19,277 houses inspected, the plumbing was defective in 5,944, there being 5,357 leaky faucets. There were 9,275 night inspections of house-drains made, which showed that in 940 of them water was wasted at the rate of one to five gallons a minute. At this rate, there are about 10,000 houses in this city which deliberately waste about 30,000,000 gallons of water daily. That is to say, ten per cent. of the water-takers use their own proper proportion, and throw away three times as much as they use. Or to put it another way, ten per cent. of the consumers make way with nearly forty per cent. of the water distributed for equal use by all. If these wasteful consumers were restricted to their proper proportion of the water, each one of the legitimate users might have nearly fifty per cent. more water than he now has, without increasing the supply to the city. Or the number of consumers might be increased forty per cent. without increasing the supply.

"Now, the water-takers last year numbered about 97,000. Of these, 11,000 were supplied by meters, and used 20,000,000 gallons a day. The remaining 86,000 used 85,000,000 gallons a day and paid $1,367,000 for it. Judging from the result of the night inspections, it would appear that of these, 77,000 used about 58,000,000 gallons a day, and paid about $16 a year each, or $58 per million gallons. The remaining 9,000 made away with 27,000,000 gallons a day, and paid $16 a year each, or about $15 per million gallons. And yet we are told that it is not proper to compel these parties to pay the $43 per million gallons of which they defraud the city. For fraud it is, and no other name is proper for it. It is as much a crime as it would be to take trees and flowers out of the Central Park and put them in private houses or throw them away; or to take furniture out of the public buildings or paving-stones out of the streets.

"The 11,000 large consumers supplied by meter paid $80 for each million gallons they used. Why should the 9,000 domestic wasters be allowed to get their water for $15 a million gallons? Why? Only because somebody is afraid that part of the money saved will stick to some one's fingers. On that theory, all public work should cease. The suppression of waste would save at least 20,000,000 gallons a day, or enough to supply 25,000 families liberally. This would represent an income to the city of at least $400,000 a year, and would be cheaply attained by the expenditure of one-fourth of that sum."

WASTE-PREVENTION IN GREAT BRITAIN.

THE SANITARY ENGINEER had investigated to some extent the methods employed in the cities of Great Britain, and in response to the following letter, published the regulations of Liverpool and Glasgow, given below:

To the Editor of THE SANITARY ENGINEER: May, 1882.

SIR: At the recent meeting of the American Water-Works Association in Columbus, O., much interest was manifested and considerable discussion had regarding the unnecessary use and the waste of water. The general application of meters to all classes of consumers was advocated by some managers of works as the only efficient means of checking wastefulness. Others seemed inclined to the view that the real cause of the excessive consumption of water is not reached by meters. They insure payment for all water used, to be sure, and thus tend to make the consumer sparing in the use of water, but they do not repress the constant wastage of small amounts through defective plumbing, which seems to be the principal factor in producing the enormous night consumption, which is out of all proportion to the use in the day time.

As this matter has been more thoroughly investigated in some English towns than in this country, and practical results have been attained in the checking of waste, the publication here of the methods employed and the results reached there would be of great value to investigators on this side of the water, to whom the publications in the proceedings of English societies are inaccessible, and I hope that you will find space for them in your paper. Your obedient servant, J. J. R. CROES.

THE PREVENTION OF WASTE IN LIVERPOOL AND GLASGOW.

IN 1881 we addressed certain inquiries to Mr. J. Parry, engineer in charge of the Rivington Water-Works, Liverpool, on points connected with the waste-prevention system employed there, and have received the following reply:

MUNICIPAL OFFICES,
LIVERPOOL, October 10, 1881.

To the Editor of THE SANITARY ENGINEER:

In answer to your inquiries I have much pleasure in furnishing you with the following information with respect to the Liverpool water-supply.

The number of inhabited houses supplied with water from the works of the Corporation is 122,673, and the number of inhabitants is 725,000. The average number of persons to each house is therefore 5.91.

The average rate of consumption for all purposes is 23 (imperial) gallons per head per diem. This includes over 5 gallons per head per day for trade purposes measured by meters, and about 3 gallons per head per day for public sanitary purposes and for miscellaneous trade uses charged for by assessment and not sold by meter measurements, leaving about 15 gallons per head per day as the average rate of consumption for domestic purposes, shops, offices, hotels, public houses, warehouses, public buildings, and all waste.

The water is constantly laid on at a pressure sufficient to reach the top of every house in the district, and to enable fires to be extinguished by means of hose attached to hydrants fixed on the mains, and without the intervention of fire-engines. Along the line of docks, jets of water can be thrown from the fire-hydrants to a height of 80 feet above the street levels.

There is probably no modern city in which the legitimate demands for water and the facilities for using it are greater than in Liverpool. Water-closets are general; private fixed baths are common; and for many years it has been the practice to put fixed baths in all new houses exceeding about £18 in annual value; public baths and wash-houses are provided to a larger extent than in any other city in the country; there is a considerable demand for the supply of ships frequenting the port; public drinking-fountains are numerous; and water is freely used in flushing sewers and drains, and in street-sprinkling.

No extra charge is made for supplying water to water-closets or private baths.

In reply to your question as to the work done by the waste-water inspectors, I have to state that the number of first inspections for waste (that is, exclusive of examinations) made during our last statistical year (ended on the 31st day of December, 1880) was 115,602; and the number of leakages discovered from defective cocks, cisterns, and pipes was 29,521, of which 7,779 were repaired by the inspectors, free of charge, in the course of their visits.

The number of fittings that passed through our Testing and Stamping Department was 83,613, of which the testing officers stamped 71,969, the remaining 11,644 being rejected in consequence of defects revealed by the tests applied.

Concerning our system of waste-prevention, I cannot do better than to refer you to pp. 149-153 of my little book on water, of which you have a copy.

I am, dear sir, yours faithfully, J. PARRY.

In addition to the inspection by day and night, the regulation of the fittings allowed to be used plays an important part in the repression of waste in Liverpool.

LIVERPOOL CORPORATION WATER-WORKS—REGULATIONS.

4.—All fittings used in connection with a supply of water must be tested and stamped by the duly authorized officer before being fixed, and the following fees will be charged:

Bibb and Stop Taps	2d. each.
Ball-Taps	3d. "

5.—A set of standard fittings such as have been hitherto approved as exhibited in the Stamping Office; but the engineer will give due consideration to the claims of any other fitting which may be presented for approval, and which, if considered satisfactory by the committee, will be stamped, the sample purchased, and placed among and become one of the standard fittings. Before any fitting is withdrawn from among the approved samples, six months' notice will be given to the Master Plumbers' Association.

6.—Every service-pipe hereafter laid or fixed below ground shall be of lead; and every joint on every lead pipe, whether below the ground or not, shall be of the kind called a plumbing or wiped joint.

11.—Pipes of any other metal than lead shall only be fixed after samples thereof have been submitted to and approved by the Water Committee.

12.—No pipe shall be laid through, in, or into any sough, drain, ash-pit, manure-hole, or other place from which, in the event of decay or injury to such pipe, the water might be liable to become fouled or to escape without observation, or without occasioning the necessity for immediate repair.

13.—No pipe shall be brought above the level of the ground outside any building, except for the supply of an outside tap in a yard, in which case the pipe shall be properly protected from frost by brick-work, or otherwise, and encased in felt, or other non-conducting material, to the satisfaction of the engineer.

14.—Every separate service-pipe must be provided with a stop-cock and box, which will be fixed outside the private premises by, and, in case of domestic supplies within the city, at the expenses of the Corporation.

16.—Storage cisterns must be provided for all domestic supplies. Where there is a bath or hot-water apparatus, the cistern must hold not less than 50 gallons for each house. In other cases, not less than 25 gallons for each house.

17.—Cisterns for the storage of water (not including water-closet and urinal regulating cisterns) shall, if of wood, be lined with lead of not less than five pounds to the square foot. The iron, wood, or slate work shall be strong and well put together, and

APPENDIX. 51

each cistern shall be provided with a ball-tap, which must be securely fixed to the side thereof, and it must be in such a position as not to become submerged when the cistern is full, and the level of the water at such time shall be three inches below the overflow.

18.—Domestic boilers, water-closets, or urinals must in all cases be supplied from cisterns. All cisterns for the supply of water-closets or urinals shall either be on the alternating-valve principle, and so constructed as to be capable of delivering two gallons at each flush, which must be discharged within fifteen seconds, or otherwise so arranged as to produce the same result in equally efficacious manner; but no valve except the ball-cock shall at any time have a greater pressure upon it than that due to the head of water in the cistern.

24.—The pans or basins of all water-closets, not of the trough kind, must be of a semi-circular shape, or of such other form as can be most efficiently flushed; the down pipe from the cistern to the basin of the closet must be of not less than $1\frac{1}{4}$ inches in diameter, except in connection with pan-closets, where the head of water exceeds eight feet, when the down pipe may be of one inch diameter. In the case of pan-closets, the metal pan shall be capable of bearing a weight of seven pounds.

25.—The detailed arrangements of all trough-closets shall be submitted to and approved by the Engineer before such closets are fixed.

26.—Every bath must be provided with a well fitted and perfectly water-tight ground outlet-plug, with chain, complete, or such outlet-tap as shall be entirely independent of the inlet.

27.—None but screw-down taps, incapable of being suddenly closed, shall be fixed on pipes supplied direct from the main.

28.—Any stand-pipe fixed for the use of the occupants of more than one house must be fitted with a self-closing apparatus incapable of being suddenly shut.

29.—The Corporation will provide and fix all water-meters for the supply of water for trade purposes, and will also lay the service-pipe from the boundary of the premises to the inlet of the meter, and fix the stop-cock thereon at the expense of the occupier of the premises.

30.—No steam-boiler or any description of closed boiler will be allowed to be supplied direct from a service-pipe; but the supply will be given through meter, and a self-acting check-valve must in every case be fixed on the pipe so as to prevent a return of the water.

31.—Hydrants for fire or other purposes inside premises can only be permitted by the special sanction of the Water Committee, for which application must be made in every case.

32.—Before a connection for the supply of water can be made, or before any additional fittings can be connected to an existing service-pipe, the work must be inspected and approved by the proper officer of the Corporation.

33.—Printed forms will, upon application, be furnished to plumbers who have signed an agreement to conform to the regulations, which they will be required to fill up and deliver at the engineer's offices, as notices of fittings being ready for inspection, and also of any alterations made in existing service-pipes or fittings, and such notices must be given before pipes or other fittings are covered.

GLASGOW CORPORATION WATER-WORKS.

Rules to be observed with respect to the Supply of Water, and the Apparatus allowed to be used.

II.—SERVICE-PIPES.

5. Every house, in the case of self-contained houses, and every tenement, shall have its own separate service-pipe, except in the case of a group or block of houses, the water-rates for which are paid by one owner, and which are supplied by self-closing wells.

6. No house or tenement shall, unless with the sanction of the engineer, have more than one service-pipe.

7. Any premises or portion of a house or tenement occupied for trade purposes must have a separate service-pipe leading from the pipes of the commissioners to the premises or portion occupied by each tenant.

8. All the service-pipes, unless otherwise specially agreed upon, shall be of lead, and of the following weights per yard:

$\frac{1}{2}$-inch diameter, 7 pounds per lineal yard.
$\frac{3}{4}$-inch   "   10   "   "
1-inch   "   14   "   "
1$\frac{1}{4}$-inch   ".   18   "   "
1$\frac{1}{2}$-inch   "   24   "   "

Overflow-pipes may be of lighter weights.

9. When the service-pipes are allowed to be of iron, they shall be of the description, and laid and jointed in the manner approved of by the engineer. They shall be of the following thickness and weight:

Diameter of Pipe.	Thickness of Metal.	Depth of socket.	Length of Pipe, exclusive of Socket.		Mean Weight of each Pipe.		
Inches.	Inches.	Inches.	Feet.	Inches.	Cwt.	Qrs.	Lbs.
4	$\frac{3}{8}$	3	9	0	1	1	25
3	$\frac{3}{8}$	3.	9	0	1	0	17
2	$\frac{3}{8}$	3	6	0	0	2	4

10. The point of discharge of all pipes must be above ground, and visible to the occupiers of the house and the commissioners' inspectors.

11. The service-pipes must be laid at least two feet six inches below the surface of the ground, and if in exposed places must be properly protected from frost.

12. The ferrule which connects lead service-pipes with the distributing-pipes of the commissioners must be of brass, and screwed into the iron pipe. A stop-cock shall be attached to the service-pipe in the street, at a distance of about two feet from the distributing-pipe, with a cast-iron cock-box supported upon brick-work in mortar. It is recommended that another stop-cock, being a screw-down or slide-valve cock, be put on the service-pipe inside the premises, close to the outside wall, in a position easy of access, so that the water can be turned off in case of frost or accident.

13. The joints on all lead service-pipes shall be "plumbing" or "wiped" joints.

APPENDIX.

14. The connection with the commissioners' pipes will be made by the commissioners' workmen, and no other parties will be allowed to open, shut, or in any way interfere with any of the pipes, valves, or apparatus belonging to the commissioners, on any pretext whatever.

III.—CISTERNS.

15. The supply of water will be constantly laid on at the greatest pressure which the reservoirs and works of the commissioners will give; but it is necessary to have cisterns of sufficient capacity for the ordinary domestic supply in all houses and flats of houses at a greater elevation than about 200 feet above the mean level of the sea on the north side of the River Clyde, and 150 feet above the sea on the south side of the river. These elevations vary with the distance from the reservoirs from which the supply is drawn; but all necessary information will be given on application at the engineer's office.

16. Every cistern must be made and maintained perfectly water-tight, and be provided with a ball-cock, which must be branched into the inlet-pipe, and secured to the side of the cistern.

17. All cisterns must be placed in such positions that they can be easily inspected, and sufficient space must be allowed for repairs.

18. The top of the waste-pipe must be one and a half inches from the top of the cistern, and the ball-cock so adjusted that it will be shut when the water is two inches from the top of the waste-pipe, or three and a half inches from the top of the cistern, and so as not to become submerged when the cistern is full.

19. The overflow-pipe shall not be joined to any pipe or drain, but must have an open end brought to the outside of the premises, above ground, at a conspicuous point, or made to discharge overhead within the premises, that the discharge, when there is any, may be visible to the occupiers of the house, or the commissioners' inspectors or the police, so that waste of water may be prevented.

20. When there is a separate opening at the bottom of the cistern for cleansing, it must have a ground brass valve.

21. All public works to have cisterns of sufficient capacity to hold a few hours' water.

IV.—TESTING AND STAMPING.

22. All the fittings used, including all those supplied from cisterns, and including all ordinary kinds where the supply is by meter, must conform to samples approved of by the engineer, and must be tested and stamped by the proper officer of the commissioners. When a meter is removed from any premises, all unstamped water-fittings must be also removed.

23. Samples of the fittings which have been approved of may be seen in the Testing and Stamping Office, and the engineer is prepared to receive and examine any other fitting, and, if approved of, to sanction its use. A general description of the fittings approved of is given at the end of these rules.

24. The following fees will be charged for testing and stamping:—

Nose, stop, and tube cocks	2d. each.
Bath and Lavatory fittings	2d. "
Ball-cocks	3d. "
Water-closet cisterns	3d. "
" " with ball-cocks	6d. "
Stop-cock boxes	2d. "

V.—TAPS.

25. Screw-down cocks only shall be used when the supply is from the main service pipe, or from a cistern on any other than the same floor.

26. Where the supply of water is from a cistern on the same floor, to baths, wash-hand basins, etc., loaded valves, common ground cocks, or ground cocks with stuffing-boxes may be used. In self-contained houses these cocks may be supplied from a cistern in the flat above.

27. No common ground cock shall be attached to the main service-pipe, except the cock in the street.

28. Easy access for inspection shall be given to all cocks attached to wash-hand basins, baths, etc., by means of a hinged door.

VI.—WATER-CLOSETS.

29. No water-closet will be allowed to be supplied direct from a service-pipe, but must be supplied from a cistern on the same flat as the closet, and no water-closet shall be supplied by a tap of any kind. In self-contained houses the lever and crank may work to a cistern in the flat above.

30. Every cistern for the supply of a water-closet must be fitted with an efficient waste-preventing apparatus, so constructed as not to be capable of discharging more than two gallons of water at each flush, and so that it cannot be made to flow continuously either by intention or neglect. The cisterns must contain from five to eight flushes in all places where the supply of water is not constant.

31. The cisterns supplying all pan-closets must have a proper service-box attached.

VII.—URINALS.

32. Every urinal must be supplied from a cistern, which must be fitted with a waste-preventing apparatus similar to that above described for water-closets, so constructed as not to be capable of discharging more than one-half gallon of water at each flush, or supplied by a cock similarly constructed.

VIII.—BATHS AND WASH-HAND BASINS.

33. Every plunge-bath and wash-hand basin shall be so constructed that the water shall flow in above the bottom, and be discharged from the bottom.

IX.—WELLS IN COURTS.

34. All water-taps in courts and closes, and on common stairs, to be self-closing, secured in an iron case, and properly protected from injury, and so arranged that they cannot be left in such a position that the water will continue to run.

X.—BOILERS.

35. No boiler for generating steam shall be supplied with water by a direct connection with the service-pipe. Every such boiler must be supplied from a cistern either by a common feed-pump or an injector.

XI.—METERS.

36. Application for a supply of water by meter must be made on the printed form prepared for the purpose, and which can be obtained on application at the engineer's office.

37. All meters will be supplied by the commissioners, and a rent will be charged for the same, which will include all repairs, except damage by frost, or fire, or by the working of any engine propelled by the water. The position of the meter, and the line of the service-pipe leading to it, must be the subject of special agreement with the engineer.

38. Where the communication pipes are of cast-iron, the commissioners will provide all materials, and execute all the work necessary in introducing the water into the premises, up to and including the fixing of the meter, and the whole cost thereof shall be payable by the consumer.

XII.—WATER-ENGINES.

39. Air-vessels of such dimensions as may be approved of by the engineer shall be attached to all water-engines.

XIII.—WATERING-TROUGHS.

40. All watering-troughs for cattle to be of cast-iron, and fitted with a proper ball-cock in a covered division of the trough.

XIV.—ALTERATIONS ON WATER-FITTINGS.

41. Notice must be given to the engineer, in the form referred to in Rule 2, of the intention to make any alteration upon the existing service-pipes in any premises, or upon any apparatus connected therewith, or of the intention to fit up any new service-pipes or apparatus, that the same may be inspected by the proper officers of the commissioners, and consent given for the intended alterations; and such alterations must be in accordance with these rules, and the new fittings must be stamped.

42. If any lead service-pipe is less than the weight given in Rule 8, and is found out of repair and wasting water twice within six months, the whole of the same, so far as it is underground or in unoccupied cellars, must be removed and replaced by pipe of the proper weight.

Section 43 provides that defective valves, taps, and worn-out fittings shall be replaced by approved apparatus.

44. If any tap in a court or close, or on a common stair, or at a urinal, be found open and wasting water twice within three months, though it may not waste water when properly shut, the same must be removed, and replaced by a self-closing cock.

XV.—WASTE OF WATER.

45. Information of any continued rushing sound heard in a house should at once be sent to the Water Office, or communicated to the engineer; and all persons are requested to give timely information of any leakage or waste of water, whether the same be accidentally, negligently, or willfully occasioned or suffered.

JAMES M. GALE, *Engineer*.

WATER-WORKS OFFICE,
23 MILLER STREET, GLASGOW, 4th November, 1879.

WATER MAY BE CUT OFF OR TURNED OFF UNDER THE FOLLOWING CIRCUMSTANCES.

For default in payment of water-rates. (The Water-Works Clauses Act, 1847, Sec. 74.)

For not providing a proper cistern and ball-cock when required, or for not keeping the same in good repair, in districts where the water is not constantly laid on under pressure. (The Water-Works Clauses Act, 1847, Sec. 54.)

For refusing the officers of the commissioners admittance into dwelling-houses, or other premises, between the hours of 9 A. M. and 4 P. M., for the purpose of examining if there be any waste or misuse of water. (The Water-Works Clauses Act, 1847, Sec. 57.)

For not having all cisterns, closets, soil-pans, and baths, so constructed and used as effectually to prevent waste, misuse, or undue consumption of water. (The Glasgow Corporation Water-Works Act, 1855, Sec. 85.)

For wrongfully doing anything in contravention of the Glasgow Corporation Water-Works Acts, or for failing to do anything in those acts provided for the prevention of waste, misuse, undue consumption, or contamination of the water. (The Water-Works Clauses Act, 1863, Sec. 16.)

If the apparatus in any house or premises be out of repair, or be so used or contrived that the water is, or is likely to be, wasted, misused, unduly consumed, or contaminated. (The Glasgow Corporation Water-Works Act, 1847, Sec. 9.)

PENALTIES WILL BE EXACTED FOR THE FOLLOWING OFFENSES.

For suffering any cistern, pipe, ball or stop cock to be out of repair, so that water is wasted. (The Water-Works Clauses Act, 1847, Sec. 55.)

For supplying or willfully permitting any person to take water who does not make an agreement with the commissioners. (The Water-Works Clauses Act, 1847, Sec. 58.)

For taking water without agreement. (The Water-Works Clauses Act, 1847, Sec. 59 ; and the Water-Works Clauses Act, 1863, Sec. 20.)

For breaking or injuring any lock, cock, valve, pipe, work, or engine belonging to the commissioners, or for drawing off the water from any reservoir or pipe belonging to the commissioners. (The Water-Works Clauses Act, 1847, Sec. 60.)

For willfully or negligently suffering any pipe, valve, cock, cistern, bath, soil-pan, water-closet, or other apparatus to be out of repair and wasting, or likely to waste, water. (The Water-Works Clauses Act, 1863, Sec. 17.)

For using water for other purposes than those agreed upon. (The Water-Works Clauses Act, 1863, Sec. 18.)

For fixing any pipe or apparatus to the pipes belonging to the commissioners, or to a service-pipe belonging to any other person ; or for making any alteration in any such service-pipe or apparatus without the consent of the commissioners. (The Water-Works Clauses Act, 1863, Sec. 19.)

GENERAL DESCRIPTION OF WATER-FITTINGS.

The taps must not be less than of the following weights and dimensions :

	SIZE OF TAP.				
	⅜-inch.	½-inch.	⅝-inch.	¾-inch.	1-inch.
Screw-down loose-valve nose-cocks & stop-cocks.	9½ ozs.	12½ ozs.	16½ ozs.	21 ozs.	35 ozs.
Diaphragm nose-cocks and stop-cocks..........	10½ "	14 "	18 "	23 "	39 "
Double-valve nose-cocks.........	11¾ "	14½ "	18½ "	22¾ "	....
Screw-down loose-valve tube-cocks.............	.....	....	27 "	..... .	....
Diaphragm tube-cocks...........................	.....	....	28½ "	.....	....
Outside diameter of tube.....................	... ...	....	⅞ in.	....	....
Underground stop-cocks, weight.	....	24½ ozs.	....	35 ozs.	54 ozs.
Length of cocks...............................	... ...	5¾ ins.	....	5¾ ins.	6¾ ins.
Ferrules, weight...............................	....	....	....	8 ozs.	14½ ozs.
Number of threads of screw to the inch.......	....	....	....	14	11
Stop-cock boxes, 8 inches high, weight.........	....	....	28 lbs.	....	....

All taps must have the maker's name stamped upon them.

The spindles of all screw-down loose-valve cocks, the screws of all diaphragm-cocks, and the keys of all tube-cocks and underground stop-cocks must be of gun-metal.

The keys of underground stop-cocks to be cast solid.

All taps must be capable of resisting a pressure of 300 pounds on the square inch, to which they will be subjected in testing.

All ball-cocks must remain tight under a pressure of 150 pounds on the square inch with the ball half immersed.

All screws to be large and well cut, and the key and barrel of tube-cocks and stop-cocks properly ground.

In tube-cocks a passage for the water to be cut round the key, so that the raising or lowering of the tube will not affect the flow of the water.

The screws of the ferrules to be made to the gauges which may be seen in the Testing and Stamping Office.

Extract from "The Glasgow Corporation Water-Works Amendment Act, 1859":

SECTION 17. All the apparatus used or to be used for conveying water to the houses of the inhabitants and manufactories, or other premises supplied or to be supplied with water under the provisions of the recited act, shall be subject to the approval of the engineer to the commissioners; and in case of dispute between the parties providing such apparatus and such engineer, such dispute shall be determined by the Water Committee of the Commissioners, whose decision shall be final.

CISTERN *versus* VALVE SUPPLY FOR WATER-CLOSETS.

IN the early part of April, 1882, there was considerable controversy over the regulations adopted by the New York Board of Health under the plumbing law requiring that in new work water-closets should be supplied from cisterns instead of direct from the service-pipes to the valves attached to the closets. During the controversy, Prof. Charles F. Chandler, then President of the Board of Health, addressed the following letter to the Editor of THE SANITARY ENGINEER:

HEALTH DEPARTMENT, NO. 301 MOTT STREET,
NEW YORK, March 25, 1882.

The Board of Health of this city have been urged to modify Rule 12 of the Plan of Plumbing and Drainage adopted by them, to the extent that water for flushing may be taken from drinking-water mains, or from tanks containing drinking-water, direct to a valve on the water-closet; the tank to be dispensed with if a check-valve be placed on the pipe connected with the supply-valve.

That part of Rule 12 affected by the modification is as follows:

"All water-closets inside the house must be supplied with water from a special tank or cistern, the water of which is not used for any other purpose. The closets must never be supplied directly from the Croton supply-pipes. A group of closets may be supplied from one tank, if on the same floor and contiguous."

It is held that the taking of water for flushing water-closets from special cisterns is a needless expense, and that no adequate advantage to the occupants of houses is gained, and that it is a hardship to manufacturers who make valve-closets, which are intended to be used without cisterns.

The correspondence on the subject thus far received is sent to you with the request that you will give space in your columns to make it public. It is hoped that the opinions of your readers, as well as your own, will be given concerning the propriety of modifying the rule as suggested. It is the desire of the board that these suggestions should be discussed in all their bearings. They further desire to state that they wish to insist on no requirement that they can be convinced is unnecessary.

Yours very truly,   C. F. CHANDLER, President.

This brought out a large number of responses sustaining the action of the Board. A few of the letters from water-works engineers who years ago went through the struggle that efficient water-works managers in this country are now carrying on against public ignorance and misapprehension, and others which are pertinent, in connection with the consideration of the causes of water-waste, are republished here, in addition to the regulations of several cities where the inspection and testing of plumbing and fittings has secured very good results, which are given on previous pages. The regulations of these cities are printed not because the details are applicable in every instance to conditions of American cities, but because they will indicate what other communities have adopted with very great profit to themselves, and which would be a guide in adopting regulations in cities of this country. So far as the cocks and waste-preventing cisterns that are required as a standard in the cities possessing the best managed water-works in Great Britain are concerned, they are, every one of them, in all their essential details, public property, and there is no reason to suspect that the authorities of any of the cities in this country would exercise their discretion by compelling the consumer to adopt any particular fitting or appliance or method of doing work which would result in fostering a monopoly.

MR. JOSEPH P. DAVIS, who was City Engineer of the city of Boston from 1872 to 1880, wrote as follows :

To the Editor of THE SANITARY ENGINEER :

SIR : The New York Board of Health asks, through your columns, for the opinions of your readers concerning the propriety of modifying Rule 12 of the Plan of Plumbing and Drainage, adopted by it. The matter is one of such great interest that I venture to offer my views.

I am of opinion that the alteration suggested would result in evil ; it certainly would not effect an improvement.

Nearly all the large cities of the country are suffering in various ways, and are put to enormous expense on account of the great waste of water that is occasioned by defective plumbing and the use of improper fixtures. A very large percentage of the entire waste (which in many cities is at least fifty per cent. of the whole water-supply) is due to the use of valve water-closets. This consideration alone should prevent the board from modifying the rule in the manner suggested, but there are also strong reasons against it from a sanitary point of view.

Check-valves serve many excellent purposes, but they are not certain enough in their action to be relied upon as safeguards to health. When once placed in a supply-pipe they are out of sight and out of mind; their failure to act would only be discovered after the injury was done—perhaps health impaired or a life sacrificed.

In New York the danger of back-flow is greater than in most cities on account of the inadequate and varying pressure of the water-supply, which leaves the pipes in the upper parts of the buildings empty, and at some points and seasons admits of draught during the morning hours in basements only. From this arises another danger, which, under many conditions, will be obviated or mitigated by the use of tank-closets— namely, at the time the closet is used there may not be sufficient pressure in the pipes to cause a flush, and the proper cleansing will be apt to be neglected. A tank would serve as a reservoir; it would fill when the draught of water was least, and, therefore, the pressure in the pipes the greatest, and would furnish the necessary flush for the closet when the pressure was low. JOSEPH P. DAVIS.

MR. JAMES M. GALE, the Chief Engineer of the Glasgow Water-Works, wrote thus:

CORPORATION WATER-WORKS,
ENGINEER'S OFFICE, 23 MILLER STREET,
GLASGOW, April 18, 1882.

To the Editor of THE SANITARY ENGINEER :

SIR : I observe, in your issue of the 30th of March last, that the question of the supply of water to water-closets direct from the main service-pipe has been raised by the master plumbers of your city, and that you invite a discussion in your columns.

So convinced was I that the system was a bad one, and was the cause of a great waste of water, that more than twenty years ago I induced the Water Commissioners of this city to prohibit the further use of apparatus of that kind. During these years we have required every water-closet to be supplied from a cistern on the same flat as the closet, and that every such cistern shall be fitted with an apparatus of such a construction that the water cannot be made to flow continuously, either by intention or neglect. My experience during the twenty years that have elapsed since the issuing of the order has strengthened my conviction that it is a just and proper one.

The objections to a supply to a water-closet being drawn direct from the main service-pipe may be stated as follows :

First—The leather or other facing of the valve does not remain water-tight for a considerable length of time under the rough usage that a water-closet handle usually gets.

Second—The escape of water, when there is any, is not considered by the occupiers of houses as at all objectionable, as it passes through the basin, and appears to them as assisting in keeping the drains clean.

Third—If the handle is propped up, under a mistaken idea that the drains may thereby be kept clear, the water runs to waste continuously in large volume.

Fourth—The system can only be applied where the water-supply is constant during the busy hours of the day, and at the particular level at which the water-closet may be placed.

Fifth—In the event of any repairs being required which require the water to be shut off from the main, even for a short time, no water can be got to flush the closet, which then becomes a source of danger to health.

Sixth—On such an occasion, whatever foul air there may be in the basin will be sucked into the main, and, it may be, some highly objectionable matter as well.

It may be, though I do not admit it, that the ball-cock of a cistern is as likely to leak as the valve of a water-closet supplied off the main, but in the case of the cistern there is no chance of foul air or improper matter being sucked into the mains; and there we require that the overflow from every cistern must discharge in such a manner that the attention of the occupiers of the house may be drawn to the fact. We have a large number of these valve-closets still in use in Glasgow, but we are insisting upon their being gradually removed. If our inspectors find one of them wasting water twice within three months, we then issue an order for its being taken out, and apparatus with a cistern attached substituted. I have no faith whatever in the so-called check-valves, as my experience of these is that they are generally found to be out of order when they are wanted to act.

I cannot see why the plumbers in any city should have a preference for any specific class of water-fittings. If, by a proper system of inspection, they are all required to use the same class of fittings, no one can have any advantage over the other. Their object should rather be to have the most expensive and the strongest articles introduced into all the houses they fit up, as upon these they will most probably have a larger profit than upon cheaper fittings. I can understand why a plumber, or any other man who has a patent, tries to push the sale of his patented article; but it is against the interested efforts of such men that officials are appointed to protect the public.

The consumption of water in your city is so high already as to attract our attention here, and if you introduce valve-closets it is certain to go higher still.

I inclose a copy of our rules as regards water-fittings, and you may make any use you please of them. I am, sir,   Your obedient servant,

JAMES M. GALE, *M. Inst. C. E.*

DR. FRANCIS VACHER, the Medical Officer of Health of Birkenhead, England, wrote thus:

MEDICAL OFFICER'S DEPARTMENT.
MUNICIPAL OFFICES, BIRKENHEAD,
April 19, 1882.

To the Editor of THE SANITARY ENGINEER:

SIR: I have read with interest the controversy under the heading, "Cistern *vs.* Valve Supply," and have pleasure in adding my testimony to the wisdom of Rule 12.

The Sanitary Authority I serve requires that "no water shall be supplied direct from the main of the Corporation to any water-closet, but must in all cases be supplied by a waste-water preventer cistern, as provided in Regulation 10."

And under Regulation 10, that "every water-closet shall be supplied by a waste-water preventer cistern, double-valve, to discharge not more than two gallons at each flush, such as Guest & Chrimes', Lambert's, Ashcroft's, Anderson's, or other approved kind, and the delivery-pipe from the cistern to the closet-pan shall not be less than $1\frac{1}{4}$-inch internal diameter."

These regulations were passed in 1879, and previous to this I had frequent occasion to draw attention to the terrible risk incurred in flushing-cisterns from drinking-water mains. In this way, to my knowledge, fouling of drinking-water has actually taken place. Having one large cistern for the supply of a closet and for general purposes is also objectionable, and not without danger to health.

Doubtless a check-valve often does its work well, but it is unsafe to rely upon it, and not possible to injure its working efficiently even for a limited period. The coincident opening of a water-closet and the drawing of water below it in the house is not by any means merely a remote possibility (as one of your correspondents writes), but in some houses an almost daily occurence. I am, sir, yours truly,

Francis Vacher.

In 1881, we explained to Mr. Parry, engineer in charge of Liverpool Water-Works, what was meant by valve-closets in this country, the term in England having a different signification, and there designating a closet of the Bramah type as distinct from the hopper, pan, or plunger closets. The following was his reply, which bears on the question at issue :

Liverpool Water-Works, November 19, 1881.

Water-closet apparatus of the kind described in your inquiry were allowed in Liverpool many years ago, but they were found to be so wasteful, and so objectionable on sanitary grounds, that their use has long been interdicted. So far as I know there are none of the apparatus in use here now.

Instances have come under my notice in which the consumption of water in build ings has been reduced from sixty gallons per head per day to three gallons per head per day by simply improving the character of the water-closet and urinal flushing arrangements.

In water-closeted towns the legitimate consumption is higher by about twenty per cent. than in towns where water-closets are not common. The waste which they occasion in the absence of proper regulations is very great, and from the information you have given me concerning the New York closets, I conclude that they are responsible for a large proportion of your extravagant consumption of water. New York has the benefit of all the experience gained by the large cities of this country, and if your authorities are in earnest they will have no difficulty in effecting such a diminution in the waste as will remove any danger of a water famine for several years.

For the flushing of water-closets, all intricate and elaborate appliances are to be avoided. The most simple and useful contrivance is the double-valve cistern, with the working parts in brass, or brass bushed, and a good ball-cock, with a down pipe of not less than one and one-quarter inches diameter, from a chamber holding two gallons ; this gives a flush which thoroughly cleanses the basin, and effectually carries all filth through the soil-pipe into the sewer.

In New York, economy of water is not the only reason for urging the adoption of proper fittings. There are serious sanitary considerations involved. If water-closets are in direct connection with the public mains, without the intervention of cisterns, the water-supply is liable to pollution by foul gases.

If your authorities entertain any doubt as to the sources of waste or the remedies required, let them select for experiment a representative district, comprising four or five hundred houses, and without giving any intimation to the inhabitants, measure the volume of water flowing into it for two or three weeks. An approximate estimate may then be formed of the rate of consumption with existing fittings. Let the houses then be examined, all wasteful fittings removed, approved fittings substituted, and all underground leakage stopped. The effect of these measures upon the consumption in the district having been ascertained, data will be afforded by means of which the result of adopting similar measures throughout the city can be calculated.

I am, yours truly,   Joseph Parry.

Professor Henry Robinson, C. E., of London, contributed this valuable note:

One of the most fruitful causes of waste of water in houses has been found to arise from the direct flushing of the closets from the water-pipes. The rule in England is to prohibit this, and to require the insertion of a small service-box supplied by the service-pipe, and containing about two gallons of water. This being in direct communication with the valve of the closet, prevents more than a regulated amount of water being used at each flush of the closet. It has been found in practice that, without such an arrangement as this, enormous waste arises, and it cannot be too strongly urged on those who have to advise in the regulation of water-supply in large towns to adopt the most stringent rule on this point. Besides the prevention of waste, the introduction of such an appliance as the service-box reduces the pressure to a uniform head, which would not be the case were the flush to be taken direct from the service-pipe, and beyond this it is found that any leakage in connection with a service-box is attended with inconvenience to the householder, and this results in its prompt rectification, whereas a leak direct into the closet might continue without inconvenience to the inhabitants, but with a certain loss to the water company.

Glasgow may be quoted as a town where excellent rules are in force. The regulations here are that no water-closet is to be supplied direct from a service-pipe, but must be supplied from a cistern on the same flat as the closet, and no water-closet is to be supplied by a tap of any kind. Also that every cistern for the supply of a water-closet must be fitted with an efficient waste-preventing apparatus, so constructed as not to be capable of discharging more than two gallons of water at each flush, and so that it cannot be made to flow continuously, either by intention or neglect. It is further required that the cistern must contain from five to eight flushes in all places where the supply of water is not constant. Also that the cisterns supplying all pan-closets must have a proper service-box attached.

Leeds is another large town which may be referred to. Here the regulation is as follows: "Every water-closet has to be provided with a service-cistern of lead or cast-iron, with a ball-cock attached thereto, and every such service-cistern has also to be provided with a proper double-valve waste-preventer; and every such valve or valves has to be so arranged as to let down at each pull or lift of the valves a quantity of water not exceeding two gallons. Also, that no pipe by which water is supplied to any water-closet is permitted to communicate, directly or indirectly, with any part of such water-closet, or with any apparatus connected therewith, except the service-cistern thereof." In London, also, the water companies have strict rules requiring the use of waste-water preventers.

The above regulations may be regarded as being the outcome of the practical experience of English water engineers. In the water-works with which we are connected we have similar rules, and require them to be rigidly enforced. We do not consider that any alternative, such as specially devised valves, however ingenious, ought to be introduced. There exists a great amount of recorded experience in England, all pointing to the same conclusion, and all indicating the saving that has been effected by putting into force such rules as have been given.

The following extract from the report for 1879 of the Water Registrar of Boston, Mr. W. F. Davis, certainly gives good reasons for making the tax on these fixtures practically prohibitory:

"The permanent, serious, and continual causes of waste of Cochituate water are through the use of hopper water-closets; the so-called self-acting closets; urinals which are constructed for a continual run of water; the use of hand-hose for the purpose of irrigation; bad plumbing materials and bad plumbing-work; and the steady run of water which is suffered in winter time to prevent freezing.

CASE NO. I.

Where there were five hopper-closets supplied in twelve months they consumed	1,088,750 gallons.
By substituting pan-closets for these the consumption for the same length of time was reduced to	384,831 "
Amount saved	703,919 "

CASE NO. II.

Where there were three hopper-closets supplied, in twelve months they consumed	1,255,470 "
By substituting pan-closets for these the consumption for the same length of time was reduced to	19,859 "
Amount saved	1,235,611 "

CASE NO. III.

Where there was one hopper-closet supplied, in twelve months it consumed	554,780 "
By substituting a pan-closet the consumption for the same length of time was reduced to	100,572 "
Amount saved	455,208 "

CASE NO. IV.

Where there were three hopper-closets supplied, in twelve months they consumed	494,180 "
By substituting six pans for the three hoppers, for the same length of time, the consumption was reduced to	113,774 "
Amount saved	380,406 "

CASE NO. V.

Where there was one hopper-closet supplied, in twelve months it consumed	554,800 "
By substituting one self-closing closet, for the same length of time, the consumption was reduced to	79,205 "
Amount saved	475,595 "

"The result of the above five cases shows, in thirteen closets alone, a total saving of 3,249,739 gallons a year, or a daily saving of 685 gallons for each closet, at the same time affording all the needed service. In these cases meters are attached, and the water is doubtless shut off at night, showing, in part, that the great waste was in the working hours of the day. But for the meter, which compels the consumer to pay for all the water wasted as well as used, the estimate of loss above given would be more than doubled. Now, take the whole number of hopper-closets—*i. e.*, 16,137—and assume what experience has shown to be within the actual fact—namely, that one closet in five is wasting water in the same ratio of the five cases cited—and the total waste will exhibit the amazing aggregate of 4,419,620 gallons in every twenty-four hours."

Subsequent to the controversy over Regulation 12 of the New York Health Department, the Department of Public Works of New York, with a view to discourage the use of water-closets so constructed as to be wasteful of water, revised the water-rates as follows, though we believe these charges have not been collected in all cases:

WATER-CLOSET RATES.

"For hoppers, of any form, when water is supplied direct from the Croton supply, through any form of the so-called single or double valves, hopper-cocks, stop-cocks, self-closing cocks, or any valve or cock of any description attached to the closet, each, per year, twenty dollars.

"For any pan-closet, or any of the forms of valve, plunger, or other water-closet not before mentioned, supplied with water as above described, per year, ten dollars.

"For any form of hopper, or water-closet, supplied from the ordinary style of cistern fitted with ball-cock, and overflow-pipe that communicates with the pipe to the water-closet, so that overflow will run into the hopper or water-closet, when ball-cock is defective, or from which an unlimited amount of water can be drawn by holding up the handle, per year each, five dollars.

"For any form of hopper or water-closet, supplied from any of the forms of waste-preventing cisterns that are approved by the Engineer of ₜthe Croton Aqueduct, which are so constructed that not more than three gallons of water can be drawn at each lift of the handle, or depression of the seat, if such cisterns are provided with an overflow-pipe—such overflow-pipe not to connect with the water-closet, but to be carried like a safe-waste, as provided by the Board of Health Regulations—per year, two dollars.

"Cisterns answering this description can be seen at this Department."

The Sanitary Engineer thus commented editorially on these rates:

As far as they go we think they hit the nail on the head in discriminating against such forms of fixtures as are conceded to be wasteful. Boston, some years ago, forbade the use of the form that is to be taxed $20 per annum, on account of its great wastefulness; though there they mistakenly went to the other extreme, by insisting on a self-closing-cock, which had to be held open to permit even the most inefficient flush; consequently, such closets never were flushed and were most offensive.

APPENDIX.

Thousands of these closets are now in use in this city, cheapness being their only recommendation. It is quite time that those who, to save the cost of putting in a proper appliance, thereby cause a waste that deprives others of water that they should have, should be made to pay a sufficient tax to make this course no longer profitable.

Those taxed ten dollars per year are the better forms of water-closets usually located above the basement floor. They are also wasteful, although not to so great a degree as the former, and even when their valves are supplied from large tanks near the roof, a great deal of water has to be pumped that has leaked away through these valves without the knowledge of the owners of the building, who pay the meter charges without knowing why they are so high or how it is possible to reduce them.

The third class, taxed five dollars, are any forms of cistern-closets in which the entire contents of the cistern (if over three gallons) can be drawn off if the handle is propped up, and in which the overflow-pipe communicates with the service-box, so that if the ball-cock leaks the leakage runs unnoticed into the bowl of the water-closet. There is a probability of waste here, but not to so great a degree as in the two former kinds.

The last form, on which no extra tax is placed (two dollars being the rate hitherto), prescribes what is required in all the best managed water-works in Great Britain.

Cisterns which meet these requirements can be made by any one, and are not patentable; indeed, many involving this principle are now used over hoppers, and to come within all the requirements it is only necessary to change the overflows.

In setting this standard of what is regarded as the least wasteful, the Department have acted in accordance with what is deemed the best practice in the light of past experience.

Manufacturers of water-closets will do well to at once arrange to supply tanks that will meet this last requirement, as sooner or later there will be a demand for them. Architects will also consult their client's interest when they make specifications for new work, if they take into consideration the advantages, *pro* and *con*, of the classes of fixtures that are to be taxed annually—$5 and $2, respectively.

THE BOARD OF HEALTH RESOLUTIONS AS TO THE WATER-RATES.

At this time the following resolutions were passed by the Board of Health of New York:

"*Whereas*, It is important for the protection of this city that a water-pressure should be maintained at the highest possible level, in order to secure a supply for flushing water-closets and other apparatus; and

"*Whereas*, The diminished pressure caused by steadily increasing consumption and waste prevents the flow of water to the level of many floors, which would otherwise be abundantly supplied; therefore, be it

"*Resolved*, That the Board of Health heartily approves the recent action of the Department of Public Works, in adopting a revised scale of water-rates, which, if rightly enforced, will discourage thoughtless waste and the use of appliances which are known to be wasteful and which should be prohibited on account of the limited supply of water to this city.

"*Resolved*, That the Department of Health will co-operate efficiently with the Department of Public Works in the enforcement of all proper measures having in view the prevention of the waste of water."

THE following extracts explain more fully the matters referred to in the foregoing pages :

WATER-CLOSETS, VALVES, AND CISTERNS.

(*From the Sanitary Engineer, May* 18, 1882.)

To the Editor of THE SANITARY ENGINEER :

For the instruction and benefit of your numerous readers who do not "belong to the trade," will you not explain what is meant by the terms *single or double-valves, hopper-cocks, stop-cocks, self-closing-cocks?*" What description of water-closets are included in the ten-dollar rate? What description of water-closets are included in the five-dollar rate? What are the waste-preventing cisterns mentioned in the two-dollar rate?

By giving a plain explanation of these things, you will greatly oblige me, and, I am sure, many others who do not understand the distinctions named. Very respectfully,

P. F. VAN EVEREN,

May 11, 1882. 116 Nassau St., N. Y.

The hopper-closets to be taxed $20 per year are the ordinary enameled-iron hoppers, fitted with or connected to the valves and cocks which control their water-supply. These are innumerable in form and description. They are illustrated in every plumbing material catalogue and can be explained by any plumber. Those most commonly used in this city are a spring-valve attached to the side of the hopper and opened by the depression of the seat bearing against the spindle of the valve. They are supposed to close when the seat is relieved of weight and are aided in such action by a spring inside the valve. In some styles, called the double-valve, the water is supposed not to run when the seat is occupied, but when the pressure is removed the valve is arranged to deliver a certain amount of water and then cease running ; great numbers of valves are also set in the ground, a rod communicating from them to the seat. Another kind yet in use in this city is a stop-cock with a rod attached which extends up through the side of the seat and is operated by a crank after the manner of the ordinary wooden hydrant.

The location of hoppers of the kind described is generally where they get rough usage ; they are often exposed to frost, and being in basements and cellars, where the pressure is the greatest, they naturally leak the most water. It is a common occurrence to find the crank-cocks left open by careless servants and others, and when the spring-valves are used bricks are frequently laid on the seat to keep them running in the winter time.

The class of closets taxed $10 per year are any kind other than these hoppers, which are supplied with water through the valves described, or any other kind of valves, without the intervention of a special cistern for the water-closet. A large tank in which the water connects directly with one of these valves or cocks on a water-closet, so that the supply can be drawn without operating any valves in said tank, as, for instance, would be the case with the large storage-tanks on the upper floors of buildings in this city, would not be considered a special tank within the meaning of the rules.

APPENDIX. 67

The tax on valves and cocks on the pan and improved forms of closets is made less than in the case of hoppers, because, we presume, the conditions of their use make them less wasteful, but that they are also wasteful there is abundant evidence to prove.

For an explanation of the forms of cistern-supply taxed $5 and $2, respectively, we refer to an article by "Sanitas," quoted below, which fully explains them.

We have not illustrated any form of water-closet or valve, for the reason that it might be deemed a reflection on the particular make selected for illustration. This we desire to avoid, as the rules affect all makers and kinds alike.

THE following abstract from the articles by "Sanitas," on "Plumbing Practice," in THE SANITARY ENGINEER, explains the revised water-rates of the New York Department of Public Works:

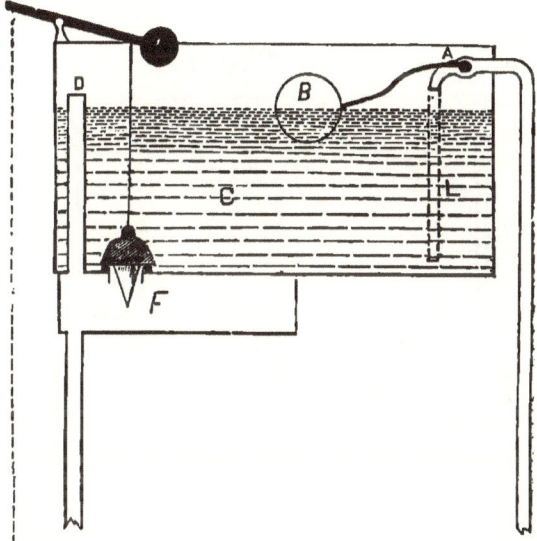

FIGURE 1.

Figure 1 shows the ordinary cistern now in use over water-closets where the flush is produced by the lifting of the water-closet handle.

This kind of a flushing apparatus is taxed $5 per year, I presume because there is a possibility of waste from two causes: First, should the ball-cock A leak, the water leaked will flow unnoticed down the pipe D into the water-closet. This pipe, it will be seen, serves as an overflow as well as an air-pipe for the service-box F. Secondly, if persons are so disposed, they can fasten up the water-closet handle so that the water can run from the cistern into the water-closet indefinitely; which is possible, because when the water lowers in the cistern, the float B will drop, which action opens the ball-cock A, so that after the contents of the cistern are drawn off then the supply from the ball-cock can continue to run out of the cistern into the service-box F, thence into the water-closet. It is true that several inventors have applied for, and possibly have obtained, patents for arranging a water-closet handle so that it can be fastened up, expecting to sell them to the unthinking people (and there are many such) who imagine that by permitting a stream of water to run into a water-closet they will keep out sewer-gas. I do not believe these devices have met with much sale, nor do I believe that with a good flush from a properly constructed cistern people will, as a rule, hold the handle up so long as to materially increase the consumption of water. I notice, however, that English engineers have acted on that assumption, and probably their experience is of more value in coming to a conclusion than my opinions. The requirement, however, to carry the overflow down like a safe-waste and not into the water-closet as hitherto is a very poor one, and needs no argument, for it seems perfectly clear that when the water from a leaky ball-cock can flow down an overflow-pipe into a water-closet such leaks are not likely to attract sufficient attention to induce a householder to have the ball-cock repaired, but when carried like a safe-waste it will be likely to receive attention.

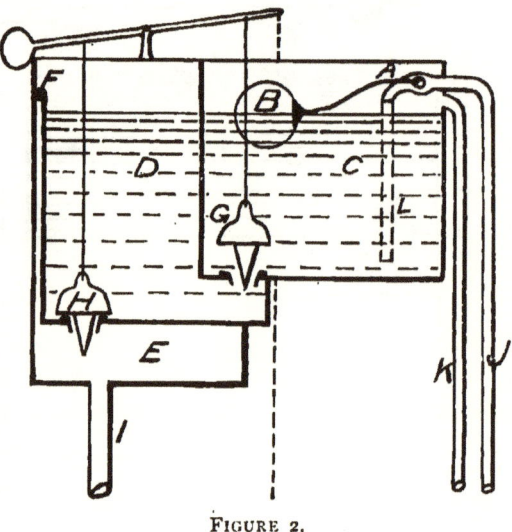

FIGURE 2.

Figure 2 shows a cistern that combines the features and meets the conditions required by the new rates to secure the lowest tax. Cisterns of substantially this character are used in most English cities. *They are not patented there, and no valid patent could be procured for them here.* Of course, some maker might get up some little detail on which he might secure a patent, but all the essential features are public property. By reference to the sketch it will be seen that when the cistern is at rest the water is at the same level in both compartments. This is possible because the valve G is lifted, permitting the water in the storage part C to flow into the measuring compartment D. When the handle is pulled the valve G closes the communication with the measuring compartment D, and the valve H is lifted, allowing the contents (3 gallons) to pass into the service-box E and thence to the water-closet. Where hoppers are used, the service-box E is not needed, but with most forms of water-closets it is necessary in order to secure what is called the after-wash, which fills the bowl after the flap-valve, pan, or plunger in the closet is restored to its place. This after-flush is possible because the valve H is larger than the outlet I to the water-closet, so that when the cistern-valve is closed the service-box retains the water that the down-pipe I has not carried off. This quantity makes what is called the after-flush. It will also be noticed that in this cistern there is no overflow-pipe communicating with the water-closet like the pipe D in Figure 1, but the pipe K serves as an overflow and warning pipe, and is required to terminate where any leakage can be readily noticed by the occupants of the house. The tracing of these overflow-pipes to see they are properly run, I imagine, will give plenty of work to the inspectors.

The tube F is to give air to the service-box E, so the water will run off when the valve H is closed.

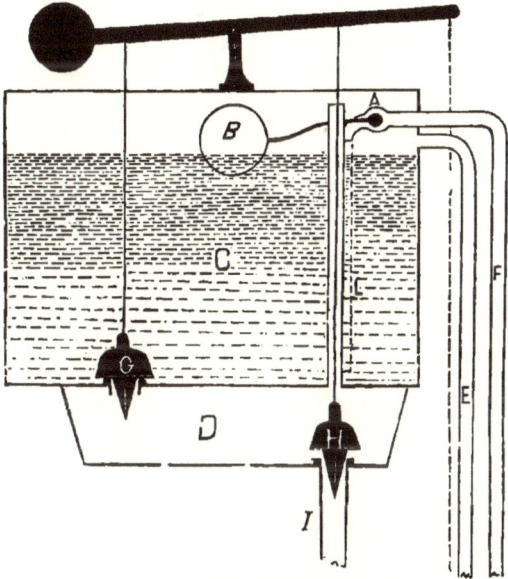

FIGURE 3.

Figure 3 shows what is familiarly called the double-valve cistern. Large numbers of them are now used over hopper-closets, and are operated by connecting a chain or wire from the lever on the cistern to a lever-attachment under the seat, which is operated when the seat is occupied, or to a door. It will be noticed that the essential features of the No. 2 cistern are also in this one, and the same result can be secured by reversing the order of hanging the valves. In this the water is all retained in the storage-compartment until the seat is depressed, then the valve G is opened, and the outlet-valve H to the water-closet is closed. No matter how long the seat may be occupied, no more than the contents of this compartment can be drawn off. It will be noticed, however, that, like Figure 1, the overflow-pipe communicates with the pipe to the water-closet. To secure, therefore, the benefit of the lowest rates the only change required in this cistern is to take the overflow down from the outside like a safe-waste from a point below the air-pipe, which should be extended.

I must confess that I think these rules are sound, and I only trust a system of inspection will be instituted and maintained that will secure the impartial enforcement of them, and I apprehend that if this is done there will be far less complaint than when parties who have cheerfully gone to the expense of meeting the wishes of the authorities find after a time that their friends and neighbors who may have ignored them are not molested. If all are treated alike, then manufacturers, plumbers, and householders can know what to depend on and will have no cause for complaint.

THE SANITARY ENGINEER:

A JOURNAL OF

CIVIL AND SANITARY ENGINEERING

AND

PUBLIC AND PRIVATE HYGIENE.

CONDUCTED BY HENRY C. MEYER.

"A large and flourishing weekly journal, covering the whole field of Sanitary Science and recognized as a leading authority upon the subject."—*The Nation.*

"It is gratifying to see that the public, as well as specialists, are enough interested in Sanitary matters to give substantial support to a periodical ably endeavoring to impart much needed and vital instruction."—*N. Y. Times.*

"THE SANITARY ENGINEER shows an excellent appreciation of what may be done in the field of Sanitary Engineering, and a practical ability for doing it."—*N. Y. Tribune.*

"The recognized American authority on all departments of Sanitary Engineering."—*Cincinnati Gazette.*

"It has done an excellent work in disseminating the most intelligent opinions on Sanitation."—*Springfield Republican.*

"THE SANITARY ENGINEER has done a noble work in the field it has chosen."—*Boston Herald.*

"THE SANITARY ENGINEER is, beyond question, the ablest publication of its class in this country. The papers prepared for it are from the pens of the ablest experts, and treat of specialties the discussion of which is essential to an intelligent comprehension of sanitary growth and progress."—*Memphis Appeal.*

"The leading journal of the kind published in the United States."—*Troy Times, Dec. 11, 1882.*

"The best journal of its class in this country."—*Bridgeport Standard, Dec. 30, 1882.*

"Standard authority on matters pertaining to its specialty."—*Am. Machinist.*

"It is a journal with a mission—that of raising the low standard of Sanitary Engineering which exists in the United States. Its large editorial staff includes the names of some of the best known and most practical writers in America on health subjects, and no pains or expense are spared in making the paper useful to the fullest extent."—*Engineering, London.*

"It would be impossible to point out a publication in which the department of public sanitation receives greater or more careful attention than in the journal of which we have received the fifth and sixth volumes. Whatever fault may be found on this side of the Atlantic with the newspaper press of America, it is an undoubted fact that that portion of the periodical literature of the United States which is devoted to science, occupies a most distinguished place amongst the scientific press of the globe. THE SANITARY ENGINEER stands high in this respect. In its pages the various subjects relating to public health—drainage, water supply, ventilation, heating and lighting—are most conscientiously attended to, shortcomings and abuses being fearlessly exposed, and care being taken to have all expressed opinions upon technical matters prepared or revised by specialists.—*Iron, London.*

Sold by Newsdealers, 10 cents a copy.
Subscription, $4.00 per annum, post-paid in United States and Canada.
Great Britain, 20 shillings.

No. 140 WILLIAM STREET, NEW YORK.
92 and 93 FLEET ST., LONDON.

THE FOURTH VOLUME

OF

THE SANITARY ENGINEER.

A VALUABLE VOLUME.

No book of more permanent value can be put into a library than the fourth volume of THE SANITARY ENGINEER. Its contents cover a great variety of topics, relating to preservation of health in our cities and homes, by some of the very best authorities in this country and England. From the great number of articles of permanent value, we select the following as indicative of the scope and character of the whole:

Report of the Committee of Award on the Essays in the Food Adulteration Competition.—Together with the text of the essays. These form a very valuable body of matter relating to the regulation of adulteration by law, and were largely instrumental in determining the character of the legislation in this State and New Jersey, and that now before Congress.

Recent History of Electric Lighting, and notes on the Present Condition of the Electric Light.—By President Henry Morton, of the Stevens Institute of Technology.

Letters to a Young Architect on Ventilation and Heating.—A series of papers on the heating and ventilation of buildings. By Dr. John S. Billings, of the United States Army.

Water Analysis.—By Dr. Charles Smart, U. S. A. A very valuable series of papers on the sanitary effect of impure water supplies, and the approved methods of water analysis.

Plumbing Practice.—By "Sanitas." A series of practical papers of general interest on good and bad plumbing in our houses.

A series of Miscellaneous Papers in Sanitary Engineering.—With special attention to the latest results attained in the Disposal of Sewage. By Prof. Henry Robinson, C. E., of London.

A very thorough discussion of the Separate System of Sewerage.—Especially as applied at Memphis. Participated in by Robert Rawlinson, C. E., C. B., of London; Mr. E. S. Philbrick, C. E., of Boston; Col. George F. Waring, Jr., and other well-known engineers.

In the Food and Drug Department will be found the text of the New York Law, the Adulteration Legislation of other States, and notes on current adulterations by experts in this country, England and France.

Among its miscellaneous articles are the text of the Plumbing Law for New York and Brooklyn; the Regulations and Blank Forms of the Board of Health adopted in pursuance of the law, and a discussion of the Regulations by Practical Plumbers in the United States and Canada.

A great variety of questions, arising in the plumbing of houses, sewering of cities, heating of buildings, and questions connected with infectious diseases are answered in its correspondence. These answers contain the very best accessible opinions on a very great variety of practical matters, and are a most valuable feature of this journal.

In its Gas and Electric Lighting Department will be found in compact form the latest discoveries and inventions in gas making, and electric lighting, and items on the current news in these industries, prepared by an expert on these topics.

The latest discoveries in the treatment or causation of epidemic and other diseases are discussed in its Sanitary and Health Notes by the best known sanitarians.

Bound in cloth, $5. *Postage,* 40 *cents.*

THE SANITARY ENGINEER,
140 William Street,
New York.

Obtainable at London Office, 92 and 93 Fleet Street, for 20s.

THE FIFTH VOLUME
OF
THE SANITARY ENGINEER

Comprises the twenty-six weekly issues from December to May, 1881-2, and it is the first completed volume of THE SANITARY ENGINEER as a weekly. As our first and second volumes are out of print, this volume, in the main, is a treatise on the elementary as well as the higher branches of sanitation.

We give below a general outline of the main subjects embraced in Volume V., subdivided, so that it may conveniently be seen why this volume should be in the hands of all, and particularly be read:

By PLUMBERS, because of the thorough discussion on cistern and direct supply for water-closets, by authorities on water-works both here and abroad, and letters on same subject from practical plumbers throughout the United States.

Descriptions of the plumbing work of the Vanderbilt houses in New York, and other fine jobs (illustrated), with practical comments on same by various plumbers.

Board of Health rules, etc., as relating to plumbing and the plumbing laws in this State and elsewhere, with discussions.

A series of articles on Plumbing Practice by "Sanitas." Illustrated.

And a variety of questions from many practical men in the trade, and the answers thereto, with illustrations.

By STEAM FITTERS, because of a series of papers by "Thermus," a practical man, on piping of buildings and subjects of actual experience. Illustrated.

By GAS FITTERS, because of the theoretical and practical questions and answers on pipes, burners and gas as an illuminant. Illustrated.

By ENGINEERS, because of a course of papers on the drainage of Washington, D. C.

The separate system of sewerage, by Robert Rawlinson.

A number of editorials on New York's water supply, and a collection of articles on water waste, giving the views of home and foreign engineers, together with reports and regulations of many cities where this subject has been thoroughly studied.

By ARCHITECTS, because of the reasons above mentioned, and on account of "Letters to a Young Architect," by Dr. John S. Billings, U. S. A. Also for the multitude of various subjects discussed and suggested of importance in building, many of which are illustrated.

By ELECTRICIANS AND GAS ENGINEERS, because of the matter of interest to gas fitters, as stated, and for information of the latest discoveries and inventions in gas-making and electric-lighting. Illustrated.

By DRUGGISTS, CHEMISTS AND PHARMACISTS, because of the report of the experts appointed by the New York State Board of Health on food and drug adulteration.

A series of articles on drug adulteration, by Alfred Senier, M. D., F. C. S., and numerous communications and notes from experts on food and drugs.

By PHYSICIANS, because of the information and comments on the new discoveries of the causes of disease, and the treatment.

And for the same reasons as architects and druggists.

By HEALTH BOARDS, for the same reasons as plumbers, engineers and physicians.

By THE HOUSEHOLDER and general public for all the reasons above mentioned.

Bound in cloth, $3. *Postage*, 40 *cents*.

THE SANITARY ENGINEER,
140 William Street,
New York.

Obtainable at London Office, 92 and 93 Fleet Street, for 15s.

THE SIXTH VOLUME

OF

THE SANITARY ENGINEER

Includes the twenty-seven issues from June 1 to November 30, 1882.

The following are a few of the many articles of permanent interest:

Illustrated descriptions of the plumbing work in the well-known residences of Messrs. William K. Vanderbilt, Cornelius Vanderbilt, Ex-Gov. Samuel J. Tilden, John Sloane. Also of the "Mills Building" (the largest office building in New York), and Minot, Hooper & Co.'s Commercial Buildings. A Russian Bath Establishment in Boston and the St. James Hotel in Baltimore.

"Berlin Sewerage." An elaborate report of the system just completed, with illustrations. By M. A. Durand-Claye, Engineer-in-Chief of Bridges and Roads of Paris.

"Trap Syphonage." Report with illustration of experiments for National Board of Health. By Edw. S. Philbrick and E. W. Bowditch. Also by Col. Geo. E. Waring, Jr. Also in England by S. S. Hellyer, Esq. Interesting and valuable data.

"Coverings for Steam Pipes." Results of Some Experiments. By Mr. M. J. Bird.

Weekly "Mortality Returns" of the leading cities in the United States, with "Notes and Abstracts" from Reports of Health Officers and correspondents of the National Board of Health. These have been printed each week, beginning with issue of October 12.

"Tin in Canned Fruits." By Prof. Albert B. Prescott.

"Negro Mortality of Memphis." By Dr. G. B. Thornton.

"Pullman Sewerage." Abstract of paper describing it. By Benezette Williams, C. E., Engineer-in-charge.

"Letters to a Young Architect on Heating and Ventilation." By Dr. J. S. Billings, Surgeon, U. S. A.

A model plan for an "Improved Tenement for Working Men in New York."

"Burning of Town Refuse at Leeds." A paper by Chas. Slagg, Assistant C. E.

"The Relative Uses of Gas and Electricity." By C. William Siemens, F. R. S.

"The Management of Heating Apparatus." By Edw. S. Philbrick, C. E.

An extract from a paper by S. S. Hellyer on the "Unsealing of Traps by Momentum"—interesting in connection with experiments made by Messrs. Philbrick and Bowditch, and Col. George E. Waring, Jr.

"Water Waste Prevention." "Regulation of Exeter Water Works." A paper by Geo. I. Deacon, M. Inst. C. E., read before Society of Arts.

"Report of Boston Committee," and report of results in Cincinnati.

"The Massachusetts Law to Prevent the Adulteration of Food and Drugs."

Practical articles on "Steam Fitting and Steam Heating." By "Thermus."

An illustrated article describing the "Accommodation for the Foreign Cattle Trade" at the Port of Liverpool. By Francis Vacher, M. D.

Plan of "Flush Tank for Country Residences."

"Text of the Illuminating Oils Bill for New York State."

"The Effect of Water Level on Disease." By Baldwin Latham, M. Inst. C. E.

"A Form of Specification for Steam Fitting of an Eight-Story Apartment House."

"Report on Oil Testers."

An article on the "Sewerage of Brighton, England." With reports by Sir Joseph Bazalgette and J. Bailey-Denton.

A series of papers on "Water Analysis," from the *Analyst*.

"Lead Poisoning." By Dr. Edw. S. Wood, Harvard Medical College.

In addition to the foregoing there are answers to a great variety of questions on the various topics treated by THE SANITARY ENGINEER, the usual record of patents granted and the weekly reviews of questions of current interest to sanitarians, and of the condition of the Gas and Electric Light Industries. This latter feature makes the paper of real value to inventors, since the aim is to state the facts without bias for either interest.

Bound in cloth, $3. *Postage*, 40 *cents*.

THE SANITARY ENGINEER.
140 William Street,
New York.

Obtainable at London Office, 92 and 93 Fleet Street, for 15*s*.

THE SEVENTH VOLUME

OF

THE SANITARY ENGINEER

Includes the twenty-six weekly issues from December 7, 1882, to May 31, 1883.

Among the articles of permanent value may be mentioned:

Letters to a Young Architect on Heating and Ventilation. By Dr. J. S. Billings, U. S. A.

Steam Fitting and Steam Heating. By "Thermus." A series. (Illustrated.)

The Edison System of Wiring Buildings for the Electric Light. (Illustrated.)

Illustrated descriptions of the sanitary arrangements in the residence of Cornelius Vanderbilt, Esq., the Berkshire Apartment House, Home for Aged Females, and the Duncan Office Building.

The Steam Heating Companies in New York. Illustrated description of.

Full abstract, with illustrations, of the records in the *McCloskey Patent Suit* for Trap Ventilation.

The New York Water Supply. A series of articles on the suppression of waste of water, giving the experience of European cities in attempting to deal with this problem, the practice now in vogue there, and the situation in American cities. These articles will be found of great value to water-works authorities and all who are interested in this question.

A discussion of the various projects for increasing the water supply of New York, including the Croton Aqueduct scheme, appears in almost every number in this volume.

Atlantic Coast Resorts. A Report by E. W. Bowditch, C. E., to the National Board of Health.

National Board of Health, Congressional Debates on the.

How the Plumbing Law is enforced in New York. A description of the methods employed by the department.

Germs and Epidemics. By Dr. John S. Billings, U. S. A.

Malaria (a series). By George M. Sternberg, Surgeon U. S. Army.

Lead Burning. Apparatus and Process described.

Gas Fitting in an Office Building. Description of work in the Mills Building.

American Practice in Warming Buildings by Steam. By the late Robert Briggs, M. Inst. C. E. A paper read before the institution.

There is also the current information of the operation of the food adulteration laws; record of rulings and prosecutions, and copies of laws. The weekly and monthly mortality table of the principal cities of the United States, together with a large amount of home and foreign health notes. The most complete collection of data on this subject published. Carefully prepared reviews of the reports of health officers, and the current sanitary literature. Answers to a great variety of practical questions on plumbing, heating, water supply, and steam fitting. Record of Patents, and the current record of projected buildings and construction notes, which includes information of special interest to Contractors, Engineers and Architects.

Bound in Cloth, $3. Postage, 40 cents.

THE SANITARY ENGINEER,

140 William Street,

New York.

Obtainable at London Office, 92 and 93 Fleet Street, for 15*s*.

THE EIGHTH VOLUME

OF

THE SANITARY ENGINEER

Comprises the twenty-six weekly issues from June 7 to November 29, 1883, and is replete with interesting information for every intelligent person.

Among the many important articles of permanent interest, the following may be mentioned :

Vital Statistics.—By Dr. John S. Billings, Surgeon U. S. Army. A series of original and suggestive papers.

Steam-Fitting and Steam-Heating.—By "Thermus." A series of illustrated articles on modern practice in the fitting of buildings with Steam Apparatus.

Letters to a Young Architect on Heating and Ventilation.—By John S. Billings, Surgeon U. S. Army. A continuation of the Illustrated series.

The Use of Lead for Conveying and Storing Water.—By Prof. Wm. Ripley Nichols. A valuable contribution giving the results of experience and investigation up to the present time.

English Plumbing Practice.—By a Journeyman Plumber. A series of illustrated practical articles of special interest to practical workers.

Model Stables.—Giving illustrated descriptions of the ventilation and drainage of the stables of Mr. Wm. Pickhardt and Mr. Frank Work, New York.

A Series of Illustrated Articles.—By F. B. Brock, giving the expired patents on water-closets, and radiators used in steam heating. Of value to those interested in the manufacture of these appliances.

Illustrated Description of the Sewerage and Water-Supply of Bunzlau, in Silesia, in 1773.—By W. Doerich, C.E. Of historical interest.

The Liernur System of Sewerage.—Carefully prepared reviews of its claim in connection with a report on the system of Dr. Overbeek de Meijer, and a discussion of its applicability for Baltimore.

The Relation of Soils to Health.—Giving results of experiments on filtering capacity of soils. By Raphael Pumpelly.

American Practice in warming Buildings by Steam.—By the late Robt. Briggs, M. Inst. C.E.

Illustrated Description of the Plumbing, Heating, Lighting, and Ventilation in the following Buildings: The Marquand Houses, Apartment house at Madison Avenue and Thirtieth Street; New Library Building, Columbia College; the Duncan-Office Building, Hawthorne Apartment House, all in New York City. Plumbing in residence of Thos. Craig, Esq., Montreal; The Government Printing-office, Washington; The Holborn Restaurant, London.

A Novel and Ornamental Fire-Escape.—Illustrated.

Plan of Improved Tenements for Working People erected by the Corporation of Trinity Church.

Plan of a Public Shower-Bath in Berlin.

The Turco-Russian Baths of Astor Place, New York.—Illustrated description.

The Vienna Electrical Exhibition.—A series of letters by an expert describing features of the exhibition. Illustrated.

There are also carefully-prepared reviews of the reports of Health Officials, Water Boards, City Engineer, and the current literature on the subjects treated by THE SANITARY ENGINEER.

Also the current information of the operation of the food adulteration laws, record of rulings and prosecutions, and copies of laws; the weekly and monthly mortality table of the principal cities of the United States, together with a large amount of home and foreign health notes, the most complete collection of data on this subject published; answers to a great variety of practical questions on plumbing, heating, water-supply and steam-fitting; record of patents, and the current record of projected buildings and construction notes, which includes information of special interest to contractors, engineers and architects.

Bound in cloth with Index, $3. *Postage,* 40 *cents.*

THE SANITARY ENGINEER,
140 William Street,
New York.

Obtainable at London Office, 92 and 93 Fleet Street, for 15*s*.

THE NINTH VOLUME
OF
THE SANITARY ENGINEER.

For Civil, Mechanical, and Sanitary Engineers, Architects, Health Officers, Plumbers, Steam-Fitters, and general readers who are aware of the rapidly increasing importance of the study of all topics affecting the public health, the Ninth Volume of THE SANITARY ENGINEER, including the 26 weekly issues from December 6, 1883, to May 29, 1884, contains much matter of great value. Its articles are prepared by the best authorities in the several departments, and are at the same time written to be understood by intelligent householders who are not themselves either Engineers or Sanitarians.

The following are among the subjects discussed in the Volume :

The Duties and Responsibilities of State and National Officers of Health, in a series of Editorials of much vigor.

Carefully-prepared Reviews of the Reports of State and Local Boards of Health, making one of the most complete records of the present condition of Sanitation in the United States and Great Britain which is accessible to the reader.

Mortality Statistics of the United States, presented in a weekly table very carefully compiled, with weekly notes on the health of the United States, Canada, and Europe.

Vital Statistics.—Several valuable papers by Dr. J. S. Billings, on the computation of these Statistics.

Numerous important articles on the Adulteration of Food.

Improved Tenement-Houses as a Business Investment.—Illustrations of Buildings in New York and London.

The Public School-Houses of New York City. —An incisive, accurate series of Reports by Special Agents, with illustrations of the most extraordinary cases.

Cottage Hospitals.—The first numbers in a series of papers by Henry C. Burdett, of London, valuable to Physicians, Architects and Sanitarians.

Lofty Buildings.—Their disadvantages. Two papers by Prof. R. Kerr, of Kings College, London. Valuable in connection with the current discussion of that subject.

Public Baths and Wash-Houses.—The first of a series of papers on the Public Provision of Bathing Facilities in Cities.

Plumbing Apprenticeship.—A discussion by Master and Journeymen Plumbers.

Its more strictly Technical Articles contain, among others: *A History of American Water-Works Practice*, in its full Reviews of Reports of Water-Works Engineers and City Engineers. These are probably the fullest notice of this subject which is accessible. Comments on Notable Examples of Water-Work Construction at home and abroad. Reports on the Quaker Bridge Dam (New York Water-Supply), by B. S. Church, C. E., and Isaac Newton, C. E.

The Water-Supply of London.—A series of papers by an English Water-Works Engineer.

Notes on Sewerage Practice in the United States and Europe.

Original Data on the Memphis Sewerage.

A thorough Description of the New Main Sewerage System of Boston, Mass.—Elaborately illustrated. These articles, both text and illustrations, were prepared by one of the Engineers in charge of the work.

Illustrated Descriptions of Plumbing, Heating, Lighting and Ventilation of Notable Buildings, showing the best modern practice. These include, among others, the Metropolitan Opera-House, Stables of Mr. Cornelius Vanderbilt, the Manhattan Storage Warehouse, the Russian and Turkish Baths in the Hoffman House, the Mutual Life Insurance Company's Building, and Bridgeport Hospital. These descriptions are prepared with great care, and are fully illustrated.

American Plumbing Practice.—By a New York Master Plumber.

English Plumbing Practice.—By an English Journeyman Plumber.

These papers show the practice of the trade in the two countries where plumbing is best developed.

The Steam-Fitting and Steam-Heating of Houses.—By a Practical Steam-Fitter, under the nom de plume " Thermus."

Gas and Electricity.—Processes of Gas Manufacture. The Vienna Electrical Exhibition is described, with illustrations, in the Special Correspondence of an American Electrical Engineer.

Healthy Foundations for Houses.—A series of papers by Glenn Brown, Architect.

Correspondence.—Containing a great variety of inquiries and replies by the best obtainable authorities on Practical Questions affecting House-Construction, Plumbing, Water-Supply, Heating, Ventilation, Sewer Building, Reservoir Construction, etc.

American Patent Records and English Patent Records.—Containing Patents granted in the department of manufacture affected by Sanitation in all its branches, Heating, Plumbing, Ventilation, etc.

Notes and Discussions on Current Topics of Interest.—Among these have been articles on the Cause of the Floods in the Ohio Valley, the Relation of Plumbers to State Medicine, Hints to Housekeepers on the Care of Mechanical Apparatus, Hygiene of Schools, etc.

Reports of Societies and Associations, Awards of Contracts, the Current Record of Buildings Projected, etc., are furnished by Special Correspondents.

THE WHOLE CONSTITUTING A VOLUME OF PRACTICAL INFORMATION OF THE HIGHEST VALUE.

Bound in cloth, with Index, $3. *Postage*, 40 *cents*.

THE SANITARY ENGINEER,
140 William Street,
New York.

Obtainable at London Office, 92 and 93 Fleet Street, for 15*s*.

THE TENTH VOLUME
OF
THE SANITARY ENGINEER

Includes the twenty-six weekly issues from June 5 to November 27, 1884.

Among the articles of permanent and special interest may be mentioned:

Notable Exhibits at the International Health Exhibition, London.—These illustrated descriptions were prepared by specialists, and possess more than usual interest, included in which are illustrations of notable lead-work, of special interest to plumbers; description of Clark's process for softening and purifying water; also illustrations and descriptions of various sewage and water filters; illustrated description of steam ovens; a history and elaborate description of the various methods of separating cream by mechanical means, and a description of refrigerating machines.

London Water Companies.—Elaborate description, extending through several numbers, of the interesting exhibits of the London Water Companies, showing section of their filter-beds, and numerous interesting details to water engineers. These papers were prepared by a well-known borough engineer, and are interesting to hydraulic engineers.

Illustrated Description of the Plumbing, Heating, Lighting, and Ventilating of Notable Buildings.—These include, among others, the new building of the Mutual Life Insurance Co., of New York; residence of Henry G. Marquand, Esq., New York; residence corner Madison Avenue and Sixty-ninth Street; Berkshire Apartment-House; and residence of H. H. Cook, Esq., New York.

Steam-Fitting and Steam-Heating.—By a practical steam-fitter, under the *nom de plume* of "Thermus." Continuation of series. Fully illustrated.

Public Urinals of Paris.—Description, with sheet of illustrations.

English Plumbing Practice.—By an English Journeyman Plumber. These articles are by a thorough workman, and of special interest to mechanics in any part of the globe.

New Method of Heating Two Boilers by One Water-Back.—With illustrations and description.

The Syphonage and Ventilation of Traps.—Criticism on the report of Messrs. Putnam and Rice on their experiments with traps, printed in the *American Architect*, and correspondence thereon.

Iron as a Material for Purifying Potable Water.—By Prof. William Ripley Nichols.

Filtration of Certain Saline Solutions through Sand.—Abstract of paper by Prof. Wm. Ripley Nichols.

Healthy Foundations for Houses.—Series of papers, illustrated, by Glenn Brown, Architect.

Rights of Tenants occupying Insanitary Houses.—Opinion of Justice Daly, of the Court of Common Pleas of New York.

Preventative Inoculation for Hydrophobia.—Comments on experiments by Prof. Pasteur.

Sewerage of Waterbury.—Description, with illustrated details.

New Orleans Quarantine Conference.—Resolutions adopted and editorial comments on the same.

Improvements in the Hull General Infirmary.—Illustrations giving plans and elevation, with descriptive matter.

Unbalanced and Lumped Bids.—An elaborate communication showing the methods adopted in France and other European countries for letting contracts for engineering and other work.

The so-called Plumbers' Trade-Protection Controversy.—A full and comprehensive history of the misunderstandings and controversy between certain plumbing societies and the manufacturers and dealers in plumbing materials in the United States during the autumn of 1884.

International Electrical Exhibition at Philadelphia.—Series of letters describing the exhibition, with illustrations.

System of Heating Houses in Germany and Austria.—Illustrated article.

Pest-Holes in New York.—Series of illustrated descriptions of some of the notable insanitary tenement-houses. Editorial comments charging the Board of Health with want of energy in dealing with these nuisances.

Cholera.—Dr. Max Von Petterkofer's views.

Public Baths and Wash-Houses.—Illustrated description of notable public baths in London.

Blunders in Plumbing.—Series of suggestive articles, with illustrations, showing the blunders made in arranging the plumbing details of houses. Notable by-passes in arranging trap-ventilation.

Heating and Ventilating Massachusetts Institute of Technology.—Description of methods employed, by S. H. Woodbridge. Fully illustrated.

Aeration of Surface-Water.—Editorial on report of Prof. Albert R. Leeds, on the results of the analysis of the Schuylkill water.

Checking the Waste of Water in Boston.—Report showing very satisfactory results due to systematic effort.

Table giving Bids in Detail for Sections A and B of the New York Aqueduct.—With editorial comments on the action of the Commissioners in rejecting these bids.

Reports by Special Correspondents—Of the proceedings of the International Congress of Hygiene at The Hague; American Public Health Association at St. Louis; the Plumbers' Congress, London; the Sanitary Institute at Dublin; National Convention of the Master Plumbers of the United States at Baltimore.

Bound in cloth, with Index, $3. Postage, 40 cents.

THE SANITARY ENGINEER,
140 William Street,
Obtainable at London Office, 92 and 93 Fleet Street, for 15s.   New York.

"*American Sanitary Engineering.*"

By EDWARD S. PHILBRICK, C. E.

Fully Illustrated with thirty-two Figures and Plans of Sewers and Sewer-Appliances, Ventilating and House-Draining Apparatus, etc.

AMERICAN SANITARY ENGINEERING, by Edward S. Philbrick, C. E., is written by a gentleman of great experience in planning sanitary works, and is especially adapted to the difficulties met with in constructing such works in climates of greatly varying temperatures. It contains a very careful summary, in brief compass, of the principles of city, suburban, and household sanitation. The subject of which it treats is generally recognized to be of steadily growing interest and importance, not only to the architect, engineer, and builder, but also to the general reader and householder, who has a vital concern in understanding the principles which secure health in his home. In this book has been presented for the first time in this country a résumé of the entire subject in a clear and convenient form for professional and non-professional men. Its value was promptly recognized and testified to by the public press, some of the notices of which we quote :

OPINIONS OF THE PRESS.

The great interests of health and life, the dangers which threaten both, and the means of preserving the one and prolonging the other, are treated in these lectures in a manner to attract public attention. There are no subjects of household or municipal economy more pressing or important than the ventilation and drainage of houses, the construction and ventilation of sewers, the drainage of towns, and other provisions for the sanitary interests of crowded cities and villages ; and Mr. Philbrick's experience as an engineer and an expert on many of these questions especially qualifies him to treat them intelligently. Every householder and every builder will find in this volume suggestions of great value.—*Boston Daily Advertiser.*

A dozen lectures covering in a peculiarly suggestive and practical manner the subjects of ventilation, house and town drainage, sewerage, and the like. The matter is presented in a way well calculated to command attention from homemakers as well as house-builders and sanitary engineers. The methods and appliances recommended have been chosen for their fitness to meet the conditions of our climate, our modes of life, and more obvious sanitary needs.—*Scientific American.*

A useful contribution to the common-sense literature of the day, and one which largely concerns the dwellers in our great municipalities, which are frequently managed on the reverse of sanitary principles.—*The Evening Mail.*

The *Sanitary Engineer* has just issued a little volume on the subject that will no doubt prove of interest to the people of all our large cities. It is a compilation of twelve lectures delivered before the School of Industrial Science at the Massachusetts Institute of Technology in 1880, and contains many valuable hints that builders would do well to take advantage of.—*New York Herald.*

This book consists of a series of lectures delivered at the Massachusetts Institute of Technology, in Boston. We are glad that the interest they awakened has led to their present publication in connected form. Not merely sanitary engineers, but all householders and dwellers in houses who are concerned with the vital questions of ventilation and sewerage, will welcome this suggestive and instructive volume. Men do not wish to be left at the mercy of builders and plumbers ; yet too often they are helpless victims, because they do not know where to go for competent and disinterested opinions concerning rival methods and devices. The literature of the subject consists largely in puffs of patent contrivances, proceeding from their inventors or vendors. Mr. Philbrick's opinions are free from this ground of suspicion, and are, moreover, based upon the condition of American society, which is not always the case with those of foreign authors.—*Engineering & Mining Journal.*

The *Lectures on American Sanitary Engineering*, recently delivered by Edward S. Philbrick before the School of Industrial Science at the Massachusetts Institute of Technology, and printed in part in the *Sanitary Engineer* and the *American Architect*, have been published in a slim octavo volume from the office of the *Sanitary Engineer*, New York, with thirty illustrations. These lectures furnish the reader, professional or unprofessional, with a very thorough and intelligent discussion of a very important subject.—*Boston Journal.*

The ventilation of buildings, the drainage of towns, and systems of sewerage receive much careful and thoughtful attention. Contains much valuable information, and should be in the hands of every householder.—*American Machinist.*

Bound in cloth, $2.00. Postage paid.

THE SANITARY ENGINEER,

140 William Street,

Obtainable at London Office, 92 and 93 Fleet Street, for 10s.   New York.

THE PRINCIPLES
OF
VENTILATION AND HEATING
AND
THEIR PRACTICAL APPLICATION.

BY

JOHN S. BILLINGS, M.D., LL.D. (Edinb.),

Surgeon U. S. Army.

PROFUSELY ILLUSTRATED.

This interesting and valuable series of papers, originally published in THE SANITARY ENGINEER, have been re-arranged and re-written, with the addition of new matter.

The volume is published in response to the general demand that these important papers should be issued in a more convenient and permanent form, and also because almost all the reliable literature on this subject has been furnished by English Authors, and written with reference to the climate of England, which is more uniform and has a higher proportion of moisture. The need of a book based upon the conditions of the American climate is therefore apparent.

The following will indicate the character of the subject-matter :

Expense of Ventilation—Difference Between "Perfect" and Ordinary Ventilation—Relations of Carbonic Acid to the Subject—Methods of Testing Ventilation.

Heat, and some of the Laws which Govern its Production and Communication—Movements of Heated Air—Movements of Air in Flues—Shapes and Sizes of Flues and Chimneys.

Amount of Air-Supply Required—Cubic Space.

Methods of Heating: Stoves, Furnaces, Fire-Places, Steam, and Hot-Water.

Scheduling for Ventilation Plans—Position of Flues and Registers—Means of Removing Dust—Moisture, and Plans for Supplying It.

Patent Systems of Ventilation and Heating—The Ruttan System—Fire-Places—Stoves.

Chimney-Caps —Ventilators—Cowls — Syphons—Forms of Inlets.

Ventilation of Halls of Audience—Fifth Avenue Presbyterian Church—The Houses of Parliament—The Hall of the House of Representatives.

Theatres—The Grand Opera-House at Vienna—The Opera-House at Frankfort-on-the-Main—The Metropolitan Opera-House, New York—The Madison Square Theatre, New York—The Criterion Theatre, London—The Academy of Music, Baltimore.

Schools.

Ventilation of Hospitals—St. Petersburgh Hospital—Hospitals for Contagious Diseases—The Barnes Hospital—The New York Hospital—The Johns Hopkins Hospital.

Forced Ventilation—Aspirating-Shafts —Gas Jets—Steam Heat for Aspiration—Prof. Trowbridge's Formulæ—Application in the Library Building of Columbia College—Ventilating-Fans—Mixing-Valves.

The book is free from unnecessary technicalities and is not burdened with scientific formulæ.

It is invaluable to Architects, Physicians, Builders, Plumbers, and those who contemplate building or remodeling their houses.

SOLD BY ALL BOOKSELLERS.

Large 8vo. Handsomely Bound in Cloth. Price $3.00, Postage Paid.

Address,   . BOOK DEPARTMENT,

THE SANITARY ENGINEER, 140 WILLIAM ST., NEW YORK.

OBTAINABLE AT LONDON OFFICE, 92 AND 93 FLEET STREET, FOR 15 SHILLINGS.

www.ingramcontent.com/pod-product-compliance
Lightning Source LLC
Chambersburg PA
CBHW020332090426
42735CB00009B/1508